Table of Contents

Page

List of Figures

List of Tables

List of Acronyms

Acronym   Definition

AFRL Air Force Research Laboratory

AFIT Air Force Institute of Technology

AWGN Additive White Gaussian Noise

COTS Commercial Off The Shelf

DHT Discrete Hilbert Transform

DRA Dimensional Reduction Analysis

DUT Device Under Test

EER Equal Error Rate

FPGA Field Programmable Gate Array

FD Frequency Domain

FVR False Verification Rate

GRLVQI Generalized Relevance Learning from Vectors Quantized Improved

HT Hilbert Transform

IC Integrated Circuits

ICS Industrial Control Systems

IF Intermediate Frequency

Acronym   Definition

IRE Intentional Radiated Emission

LLP Ladder Logic Program

LPF Low Pass Filter

MCU Microcontroller Unit

MDA/ML Multiple Discriminant Analysis Maximum Likelihood

NI National Instruments

NISAC National Infrastructure Simulation and Analysis Center

ORNL Oak Ridge National Laboratory

OSI Open Systems Interconnection

PLC Programmable Logic Controller

PMF Probability Mass Function

RF Radio Frequency

RF-DNA Radio Frequency Distinct Native Attribute

RAR Rogue Accept Rate

ROC Receiver Operating Characteristic

ROI Region Of Interest

SCADA Supervisory Control And Data Acquisition

SNR Signal to Noise Ratio

Acronym   Definition

TVR True Verification Rate

TD Time-Domain

URE Unintentional Radiated Emission

WPAFB Wright Patterson Air Force Base

PLC HARDWARE DISCRIMINATION USING RF-DNA FINGERPRINTING

I. Introduction

This chapter introduces the research topic and describes the approach taken to attain the research goals. Section 1.1 gives an overview of Supervisory Control And Data Acquisition (SCADA) systems and some of the issues and vulnerabilities pertaining to them. Section 1.2 describes the approach taken to implement the AFIT Radio Frequency - Distinct Native Attribute (RF-DNA) process relative to semi-conductor devices and unintentional emissions. Section 1.3 provides a reference for current and related research efforts.

1.1 Research Motivation

Today electronic systems are present in everyday life. It would be nearly impossible to go outside in any urban environment or any modern day office environment and not witness an electronic system of some kind. With the proliferation of Information Technology (IT) systems, large networks such as the internet, cellular phone networks, and modern television are seemingly commonplace. Less publicly discussed are the IT networks used to operate national critical infrastructure such as the networks used in nuclear power generation plants, waste water treatment, traffic grids, and sewage systems. These networks are also commonplace and have been identified as a cybersecurity vulnerability [53].

A type of system often used to control operations of national critical infrastructure is a SCADA system. SCADA systems are essentially miniature computer systems used to control industrial processes. A Programmable Logic Controller (PLC) is the most basic unit of a SCADA system and is used for controlling a particular automated process such as temperature or pressure monitoring. .

1

One of the main types of physical components in PLCs, as with virtually all electronic devices, are Integrated Circuits (IC)s. IC devices such as, Field Programmable Gate Array (FPGA)s, operational amplifiers, and microcontrollers are widely used and often manufactured overseas as a method of cost reduction. The majority of ICs used in modern military systems are made off-shore [10]. ICs can be counterfeited, or embedded with hardware trojans [1, 10].

Industrial Control Systems (ICS) can fall prey to such IC hardware vulnerabilities. A counterfeited device or a device that has been unknowingly altered, that is used in control systems for critical applications poses a significant vulnerability. Furthermore, there is in increasing reliance upon ICS networks and particularly SCADA systems to control and monitor critical process [40]. Although critical infrastructure may be owned by private companies or corporations, government also has a reliance on national critical infrastructure. This co-dependence led to the formation of National Infrastructure Simulation and Analysis Center (NISAC), a program within the Department of Homeland Security (DHS) whose mission is to research and analyze, through modelling and simulation, vulnerabilities and complexities of critical infrastructure [5, 41].

Security measures such as bit level credentials used for digital device authentication including Media Access Control (MAC) addresses and International Mobile Equipment Identity (IMEI) numbers exist as measures of security. When considering the Open Systems Interconnection (OSI) model, these measure of security are at the implemented at Application (Layer 1) or Network (Layer 5) layers. These are far from infallible and there exist methods of bypassing these layers of security [33, 51]. PLC Operating Systems often use proprietary communication protocols and are connected in vast networks. PLCs themselves have limited processing power and memory availability. Because of the nature of their implementation and operating characteristics, they are often limited in regards to defensive monitoring software such as anti-intrusion and anti-virus software.

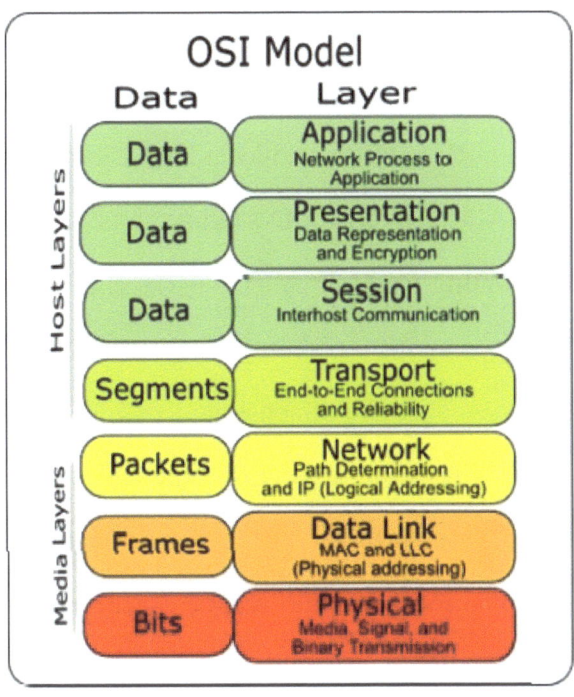

Figure 1.1: OSI 7 Layers network model [48]

Furthermore it is not uncommon for a SCADA system by remain in service for decades. For this reason they become obsolete to modern security standards and capabilities. PLC devices remain vulnerable to hardware trojans, substitutes and counterfeits.

Although work has been done at securing PLC devices at the higher layers of the OSI model, comparatively little work has been done at the lowest layer i.e. the physical waveform layer. This research augments hardware device security, in particular PLC IC devices, by means of verifying authenticity at the physical layer. While PLCs are used as a proof of concept for hardware devices discrimination, the topics contained herein apply to the majority of semi-conductor based devices.

1.2 Research Approach

The goal of this research is to use Unintentional Radiated Emission (URE) produced by IC devices as a means to discriminate between PLC devices. Inside a physical PLC device there are many points where URE may be collected. Collected emissions are taken from the microcontroller within the PLC. Previous research efforts have shown this region to be viable for collecting device URE [43]. The collected emissions are used to develop Radio Frequency Distinct Native Attribute (RF-DNA) fingerprints. The fingerprints are used to distinguish devices by exploiting Radio Frequency (RF) emission characters unique to a device that are caused by its component manufacturing variations.

Another goal is to reliably reduce the dimensionality of RF-DNA fingerprint data sets. Dimensional reduction allows for faster execution time and may mitigate adverse affects on classification performance caused by noisy, irrelevant or redundant information [3, 24]. It is expected that dimensional reduction will reduce execution time with the potential to improve classification performance.

1.2.1 Emission Collection

Using RF signal characteristics as means of device authentication as been widely researched [2, 4, 7, 9, 12, 13, 15–17, 19, 21, 36, 38, 43, 46, 49]. Although research has been done using both Intentional Radiated Emission (IRE) and URE, URE has not been as well researched. The URE signals used for device discrimination differ from IRE signals in that they are not intentionally broadcast and therefore have much lower average signal power and do not adhere to a specified broadcast pattern. IRE and URE have collection specific configurations accounting for required bandwidth and center frequency which is largely determined by the Device Under Test (DUT).

1.2.2 Fingerprint Generation

Collected signals first undergo post-collection digital processing and are then used to develop fingerprints using Air Force Institute of Technology (AFIT)'s RF-DNA process [4, 9, 36, 43]. The fingerprints are constructed from statistical attributes of the Time-Domain (TD) signal responses: amplitude, phase and frequency. The statistics used are: standard deviation (σ), variance (σ^2), skewness (γ), and kurtosis (κ). Other signal

4

features have been used in previous AFIT research such as Frequency Domain and Gabor Transform, however this research only considers TD signal responses of URE signal collections.

1.2.3 Device Classification

In *classification* a process referred to as a *classifier* uses RF-DNA fingerprints from known devices to train or develop a classification model. This model represents the known devices (Authorized Devices) fingerprint characteristics. Using the model, unknown device fingerprints are classified or aligned (correctly or incorrectly) to a particular known device represented in the model.

Device *classification* allows a one-to-many device comparison. Devices that are not represented in the model (Rogue Devices) will still be classified as one of the Authorized Devices i.e. all devices will be classified as one of the known Authorized devices. Therefore a *verification* method is used to evaluate "how much like" a device resembles a selected class.

1.2.4 Device ID Verification

Verification is a one-to-one comparison of fingerprints for an unknown device to fingerprints of a known Authorized device. The *verification* process is implemented for two scenarios: Authorized Device Identification and Rogue Device Rejection.

Authorized Device Identification examines how much like an Authorized Device looks like a different Authorized Device. Rogue Device Rejection is a comparison of how much like a rogue device resembles an Authorized Device. The intent is for the model to be able to clearly distinguish the Authorized Devices from each other, and correctly discriminate between Rogue and Authorized devices. Previous researchers have been able to use the general *verification* process using RF-DNA fingerprints to verify PLC microcontroller devices with better than 99.5% accuracy [7].

5

1.2.5 Cross Platform Validation

To validate the repeatability of the signal collection and fingerprinting process, two different collection platforms are used for signal collection. The two collection platforms are detailed in Table 3.1. The same collection method was used for both receiver platforms, as well the devices collected against, and all supporting equipment. The results from the two collection platforms are shown in Chapter 4. They are first presented independently, and are then compared directly.

1.3 Research Contributions

The research goal includes expanding upon previous AFIT fingerprinting results, and also implementing and verifying the signal collection method in [43] by replicating the process with another receiver. Previous AFIT results were expanded by examining the effects of feature dimensional reduction for both *classification* and *verification*, as well as the addition of another classifer, the Multiple Discriminant Analysis Maximum Likelihood (MDA/ML) classifier. Summarized below are the research contributions and findings related to PLC device hardware discrimination.

1.4 Document Organization

The remainder of the document is organized as follows. Chapter 2 discusses SCADA system vulnerabilities, Ladder Logic, Correlation based processing, and the *classification/verification* process using the MDA/ML and GRLVQI classifiers. Chapter 3 details the implemented signal collection process, post-collection processing, and fingerprint generation as well as feature dimensional reduction. Chapter 4 shows the results of PLC hardware discrimination using RF-DNA fingerprinting. Chapter 5 provides a summary of the findings as well as potential future work.

Table 1.1: Relational mapping between technical areas of "Previous Work" and AFIT research, and "Current Research" contributions.

Technical Area	Previous Work		Current Research
	Addressed	Ref #	Addressed #
TD Features	×	[31, 32, 36, 37] [46, 47, 49, 50]	×
SD Features	×	[7, 9, 39, 49]	
CD Features	×	[46, 47]	

Emission Type

Intentional (IRE)	×	[31, 32, 36, 37] [46, 47, 49, 50] [17, 25, 27, 28]	
Unintentional (URE)	×	[6, 7, 9, 43, 44]	×
Burst	×	[31, 32, 36, 37] [46, 47, 49, 50] [17, 25, 27, 28]	
Continuous	×	[6, 7, 9, 43, 44]	×
High SNR	×	[31, 32, 36, 37] [46, 47, 49, 50] [17, 25, 27, 28]	
Low SNR	×	[6, 7, 9, 43, 44]	×

Classification/Verification Processes

MDA/ML	×	[31, 32, 36, 37] [46, 47, 49, 50] [17, 25, 27, 28]	×
GRLVQI	×	[31, 32, 36, 37]	×
LFS	×	[25–28]	

Dimensional Reduction Analysis (DRA)

MDA/ML	×	[31, 32, 36, 37]	×
GRLVQI	×	[30, 36, 37]	×
LFS	×	[25–28]	

Verification

Electronic Components	×	[6, 7, 9, 43, 44]	×
Authorized Wireless Devices	×	[17, 36, 37]	
Rogue Wireless Devices	×	[17, 36, 37]	
Device Operations	×	[43–45]	

II. Related Work/Literature Review

This chapter gives background information on Programmable Logic Controller (PLC)s and device fingerprinting and discusses supporting research and associated academic works. Section 2.1 details the significance of PLCs in the context of Supervisory Control And Data Acquisition (SCADA) systems as well as PLC and SCADA vulnerabilities. Section 2.2 gives a description of the approach and challenges of signal collection for PLCs and details the utilization of multiple collection platforms. Section 2.3 describes PLC device *classification* and *verification*.

2.1 SCADA Systems

SCADA systems are used to automate and control large scale industrial applications such as: power generation plants, traffic grids, and waste water removal systems. They consist of a multitude of devices including PLCs and Remote Terminal Units (RTUs). Originally SCADA systems used dedicated wires for communication between devices. Although wired communications are still used today, wireless SCADA systems have become widely used, particularly in remote sensing and control environments.

The earliest SCADA systems used in the 1960s were first used in power generation plants to monitor and control sub-stations. Over the last 50 years SCADA systems have significantly evolved as computer processing power and component size continue to progress. However SCADA systems can have a lifetime on the order of decades and many legacy systems often do not have the processing capabilities and to run modern day anti-intrusion detection systems [14].

2.1.1 Programmable Logic Controllers

A particular component of a SCADA system that is used to collect sensor data and control electro-mechanical operations is a PLC, the device to which this research is focused.

PLCs are used to perform low-level operations within a SCADA system, such as sensory data input and output, and were originally designed to replace physical relays. Individual PLC devices are often referred to as modules. Modules can be specialized for certain applications e.g. power, I/O, as well as specific types of sensor modules. Some large SCADA systems (e.g. a power grid) can be comprised of hundreds if not thousands of PLCs and supporting units [22].

2.1.2 Ladder Logic Programs

PLCs perform required process using a program called a Ladder Logic Program (LLP). The name Ladder Logic originally refereed to relay logic schematics used in control and manufacturing [35]. In the advent of the digital age Ladder Logic now commonly refers to the widely used programming language used for programming PLCs. LLPs are executed by a PLC in what is called a Ladder Logic *scan*. At the beginning of a scan the PLC first reads all input values. It then performs the operations on the top-most "rung", sequentially executing all rungs. It then assigns all output values. An example LLP is depicted in Fig. 2.1.

In real world applications LLPs can be recursive and complex, containing loops and jumps. Consider a traffic light program continually looping through traffic light patterns, or the complexity of a power generation plant. However the LLPs used in this research are intentionally non-recursive, i.e. there are no internal loops. The programs themselves are very basic consisting of $N_{OP} \leq 10$ operations. This is done purposefully to ensure an experimentally repeatable signal collection process across multiple PLC devices.

2.1.3 Vulnerabilities

As previously mentioned, PLCs can have an operational lifetime of several decades. Due to their age many PLCs do not have the computer processing capability required to run modern intrusion prevention and security software. This leaves many PLCs vulnerable to

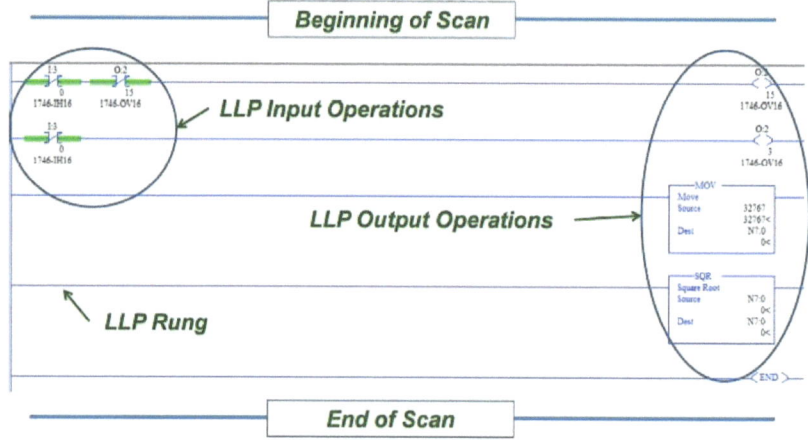

Figure 2.1: LLP example program showing a single *MOV* and *SQR* operation [43] preceded by two logic rungs.

cyber attack. A well known example of such an attack is Stuxnet, which exploited security vulnerabilities and injected malicious code into SCADA systems [53].

Extensive research has been done attempting to secure SCADA systems. Existing security measures use bit level-credentials such at the Media Access Control (MAC) address and the International Mobile Equipment Identity (IMEI) numbers to control access to a network while other software systems are used to protect against malware. However many of these measures and methods are not implemented in current SCADA systems and, in particular, PLCs. Even if implemented SCADA hardware may still be vulnerable to hardware trojans and counterfeits. An alternative to bit-level credentials has emerged using Radio Frequency (RF) radiated emissions, (unintentional or intentional) to extract unique characteristic device information at the physical waveform level that can be used to discriminate between hardware devices. This method has been shown to be succesful in a large scope of research, [4, 7, 9, 11–13, 15–20, 23, 25, 27, 38, 39, 43, 46, 49].

2.2 RF Signal Collection

2.2.1 Radiated Emissions

RF-DNA fingerprints used to discriminate among devices are constructed from captured radiated emissions from a given Device Under Test (DUT). Previous research can be categorized into two types of radiated emissions: Intentional Radiated Emission (IRE) and Unintentional Radiated Emission (URE). IRE RF energy is intentionally broadcasted and is engineered to carry information. Typically IRE RF communication signals have well defined regions such as a preamble or payload.

URE RF energy is leaked electromagnetic energy produced during DUT operation. UREs by comparison are not engineered or well structured. This creates an added challenge of repeatability when collecting URE. Radio Frequency Distinct Native Attribute (RF-DNA) IC device fingerprinting exploits characteristic differences in device waveforms caused by variances in manufactured devices. These characteristic differences are identified by calculating statistics for device's waveform attributes with the assumption that the waveforms being fingerprinted have the same structure. The IC components collected against are often shielded to mitigate RF interference to and from other components. Capturing repeated waveforms from such components requires added measures compared when to IRE collections.

URE signal collections share physical aspects with as IRE collections however URE collections often require different equipment and collection procedures. Collections can be invasive, or non-invasive. One example of a non-invasive technique is using an Electro Magnetic (EM) probe. Previous research efforts have used non-invasive techniques to capture IC electrical responses directly from connecting pins. Information exploited from this method includes (power, timing, control, data, etc.). Although these methods are non-invasive they do require contact, i.e. a physical connection is required [2, 21]. Whereas RF-based methods utilize an EM probe in close proximity to the DUT [29].

11

2.2.2 Correlation

As mentioned, this research makes use of URE which is not well structured or engineered. A method to identify an Region Of Interest (ROI) is necessary (where an ROI is analogous to a communication burst). A correlation based extraction process developed in [43] is implemented for ROI determination and extraction. The extraction process is based on a matched filter implementation which often used for the estimation of communication symbols in digital communication systems [42]. The autocorrelation ($R_{xx}[k]$) and cross-correlation ($R_{xy}[k]$) operations discussed in Section 3.3 are defined here,

$$R_{xx}[k] = \sum_{n} x_n x_{n-k}^*$$

(2.1)

$$R_{xy}[k] = \sum_{n} x_n y_{n-k}^*$$

(2.2)

Cross correlation is used in most modern day wireless communication systems as a means of signal detection. Here it is used as ROI detection. Similar to how a matched filter is implemented.

2.3 Device Discrimination

2.3.1 Classification

Using RF-DNA fingerprints *classification* is the process by which a given DUT is identified. The fingerprints from known devices are used in the classification process to develop, or train, a classification model. The established model is then used to align a DUT fingerprint to one of the known devices characterized in the classification model. One of the research goals is to use *classification* to correctly identify PLC hardware devices. Two model development processes or *classifiers* are considered and are briefly discussed in Sections 2.3.1.1-2.3.1.2.

2.3.1.1 MDA/ML

An overview of the Multiple Discriminant Analysis Maximum Likelihood (MDA/ML) process is given here, and is implemented as described in [9]. The MDA/ML process is an extension of Fisher's two class linear discriminant analysis to N_C-1 classes. The research presented here considers N_C=5 classes (5 PLC hardware devices, Authorized Devices). The MDA/ML classifier projects vectors defined by individual device fingerprints $\mathbf{F}$ using the projection matrix $\mathbf{W}$. Where $\mathbf{W}$ is the optimal projection matrix that maximizes inter-class distance, and minimizes intra-class distance.

$$F_i^W = \mathbf{W}^T\mathbf{F} \tag{2.3}$$

A device is aligned to one of the N_C classes based upon maximum likelihood conditional *posterior probability* with the assumption of equal probabilities, where likelihood is estimated for each device's projected fingerprints' assuming a multivariate Gaussian distribution [9]. Figure 2.2 shows a visual representation for a N_C-1=2 dimensional feature space, and Figure 2.3 shows the class projections resulting from projection matrices W_1 and W_2 respectively. In this example, projection matrix W_1 maximizes inter-class distance, clearly separating the three classes. Projection matrix W_2 does a poor job of separating the classes as evidence of the class overlap in the projection space.

2.3.1.2 GRLVQI

The Generalized Relevance Learning from Vectors Quantized Improved (GRLVQI) process is implemented as described in [36]. The GRLVQI process has the following advantages over the MDA/ML process: 1) there is no underlying assumption regarding the distribution of the data 2) it is well suited for situations where the number of inputs may not be consistent across classes and 3) most importantly it allows the ranking of individual features according to their ability of creating classification boundaries that minimize Bayes' risk.

13

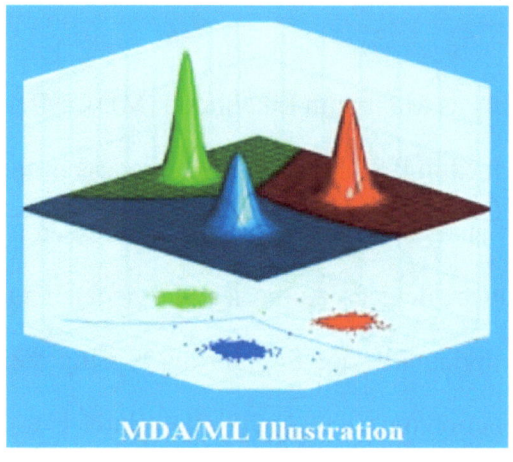

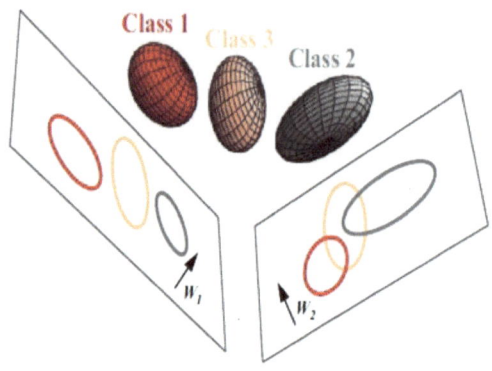

Figure 2.2: MDA/ML Model Representation for N_C= 3 Classes.

Figure 2.3: Class MDA/ML Projection onto N_C-1 Dimensional Plane

The GRLVQI process uses N_P=10 prototype vectors, where each vector is composed of N_F features, (See Table 3.3 for the feature dimensionalities considered) to represent a given device. The GRLVQI process as used in this research performs *classification* by measuring the Euclidean distance from a projected fingerprint to the prototype vectors. The projected fingerprint is classified as belonging to the class/device for which the Euclidean distance from the projection fingerprint to the prototype vector is minimized. Although other distance measures exist (Mahalanobis, Manhattan City Block, Nearest Neighbor etc.) Euclidean distance is used here, and has been shown successful in previous research [4, 34]. Figure 2.4 is a visual representation of prototype vectors representing a respective class/device with an unknown fingerprint being presented for *classification*. Figure 2.5 the shows relevance ranking for Time-Domain (TD) features for a given fingerprint.

2.3.2 *Verification*

During *classification* DUT fingerprints are aligned to a class (correctly or incorrectly). *Verification* is a one-to-one "how much like" comparison with the goal of determining weather the unknown DUT fingerprints can be verified as the known device it is being

14

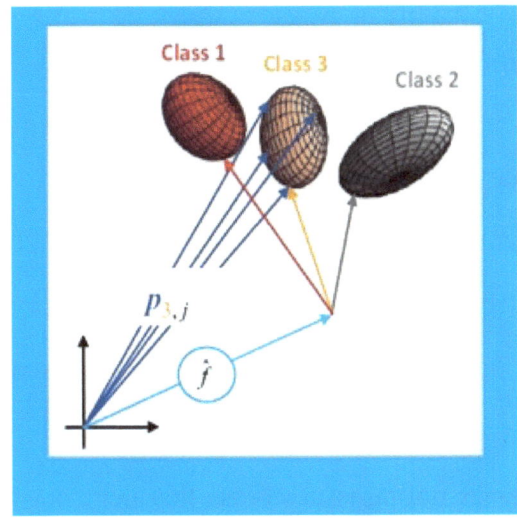

Figure 2.4: GRLVQI Feature Space

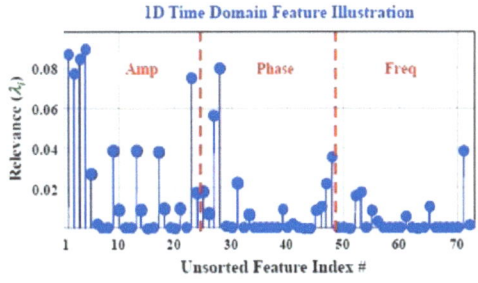

Figure 2.5: GRLVQI Relevance Rankings

aligned to during *Classification*. While a device is classified according to the class it is closest to using the selected distance metric, a device is verified (authorized or rejected) based the actual value of the distance metric.

This research follows *verification* techniques used in [6, 8, 36]. Verification results shown here are presented as Receiver Operating Characteristic (ROC) curves. Both classifiers presented in Section 2.3.1 use the same *verification* process which is further discussed in Section 3.7.2.

III. Methodology

This chapter discusses the approach taken to develop Radio Frequency Distinct Native Attribute (RF-DNA) fingerprints used for Programmable Logic Controller (PLC) device discrimination. The process is applied to data collected at Oak Ridge National Laboratory (ORNL) using the National Instruments receiver platform. The collection process is based on an existing Air Force Institute of Technology (AFIT) collection process [43]. During the signal collection phase at ORNL several receiver timing issues arose that demanded an alteration to the existing collection process. The collection alteration resulted in fewer Region Of Interest (ROI)s being collected compared to previous AFIT research. For this reason the ROI extraction method was also modified to ensure an adequate amount of ROIs. The process shown here documents the process used for collection at ORNL and reflects the changes made from previous research efforts.

Table 3.1: Receiver Collection Platforms

	AFIT Collection Platform	**ORNL Collection Platform**
Platform Manufacturer	Lecroy	National Instruments
Platform Model Number	WaveMaster	PXIe-1085 Chassis, PXIe-8135 Embedded Controller
Platform Cost	$127,000.00	$25,600.00

3.1 PLC Device Description

This research focused on applying the hardware discrimination process to 10 Allen Bradley SLC-500 PLC 5/02 CPU module devices that are collected against. The PLCs used

16

are Commercial Off The Shelf (COTS) devices whose internal Microcontroller Unit (MCU) has comparable architecture to other COTS Integrated Circuits (IC) devices [7, 9]. The devices listed are numbered/named based on variable markings and labels in the same manner as [43], shown in Table 3.2. One device, ZC was used in initial collections [43], but is not considered for comparison due to operational difficulties encountered during the ORNL collection.

Table 3.2: Component Under Test (CUT) to PLC Identity ID Mapping Based on Device Labelling and Logos [43].

Device ID	MCU Label	MCU Logo	PLC ID
Device1	NXP	None	WQ
Device2	NXP	None	WV
Device3	None	Philips	KG
Device4	None	Philips	QI
Device5	Philips	Philips	KV
Device6	Philips	Philips	OV
Device7	Philips	Philips	RG
Device8	None	Philips	ZC
Device9	None	Philips	ZZ
Device10	Signetics & Intel	Signetics	ZA

The devices are split into groups: Authorized Devices and Rogue Devices. Authorized Devices are PLC hardware devices whose RF-DNA fingerprints are used for model development which is discussed in Section 3.7. Rogue Devices are PLC hardware devices

17

whose RF-DNA fingerprints are not used during model development. Rogue Devices are used only during Verification which is further discussed in Section 3.7.2.

PLC devices are assigned as either an Authorized or Rogue device based on their relative spectral intensity plots, shown in Figure 3.1, [43]. Authorized Devices presented in this research are $\{WQ, WV, KV, RG, OV\}$. The Rogue Devices presented in this research are $\{KG, QI, ZA, ZZ\}$. Although labeled Authorized/ Rogue for the purpose of this research, all devices are authentic Allen Bradley PLCs purchased through standard COTS channels.

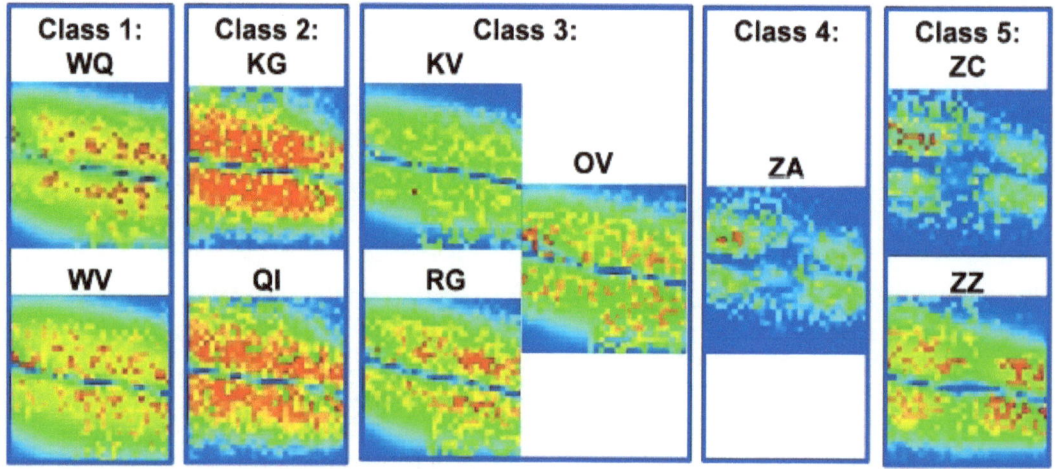

Figure 3.1: "Spectral intensity plots generated as emission maximum PSD responses over a 20×20 uniform grid above the PLC MCU surface." [43]

3.2 RF Signal Collection

3.2.1 PLC Collection Configuration

Each Device Under Test (DUT) is removed from it's manufactured housing so that the PLC mainboard is completely exposed. This allows the RF-probe to be placed on top of the MCU within a given DUT. The exposed PLC mainboard is placed onto a table which holds the RF probe and allows precise positioning of the probe in three dimensions. The

PLC mainboard is powered through extension cables with the same extension cables used for all DUTs.

There are two collection platforms used, detailed in Table 3.1. The collection platforms are configured in the same method, with one exception. Inspector software is used as the instrument controlling software for the Lecroy collection platform. Matlab® is used as the instrument controlling software for the National Instruments collection platform.

3.2.2 RF Probe Placement

Because each DUT must be connected and disconnected, the collection procedure has the potential for error in probe placement. A probe placement routine is implemented to mitigate repeatability issues. The probe placement routine developed in [43] is adopted into Matlab® (replacing Inspector) as the software to control the physical re-positioning of the probe for the National Instruments collection platform using AFIT generated control functions.

The routine has two steps: 1) Course Probe Placement - The probe is placed at a physically marked predetermined position on the DUT surface 2) Refined Probe Placement - the probe is repositioned to the site where Unintentional Radiated Emission (URE) will be collected for the purpose of generating RF-DNA fingerprints.

Once the probe has been coarsely placed based on the physical markings, emissions are collected at N_L=100 locations on a (D_X = 10) x (D_Y = 10) dimensional grid where the grid size is (x_m = 0.75cm) x (y_m = 0.75cm). At each grid location a collection is taken of the URE produced during the execution of one Ladder Logic Program (LLP) scan. During this phase of probe re-positioning the LLP being executed by the DUT is referred to as the *alignment* LLP. The *alignment* LLP consists of a known sequence of N_{OP} = 6 operations: {*MOV, SQR, MOV, SQR, MOV, SQR*}.

To determine which of the N_L=100 locations the RF probe will return to for further collection, a previously collected and stored *alignment reference* signal $x_R[n]$ representing a *pristine* alignment LLP collection is used. While there are N_L=100 alignment signal collections, there is only one *alignment reference* signal $x_R[n]$. $x_R[n]$ consists of N_{op} = 2 operations: {*MOV, SQR*}. The *alignment reference* signal is empirically chosen by means of a superior quality URE collection. The same *alignment reference* signal is used in all DUT alignment routines.

The final re-positioning of the probe (refined probe placement) is determined by cross-correlating the *alignment reference* signal $x_R[n]$ with of the each N_L=100 alignment signal collections. The location yielding the highest correlation metrics derived as in [43] is chosen as the refined probe position and all further DUT URE collections are taken at that location.

3.2.3 Sampling and Triggering

The frequency of interest for the generation of RF-DNA fingerprints is f_c=55.5 MHz. The observed clock frequency of the Allen Bradley PLC MCU is f_{clk}=18.5 MHz. The strongest component of the observed clock frequency is the third clock harmonic centered at f=55.5 MHz. To prevent aliasing during signal collection an in-line Low Pass Filter (LPF) is used with a cutoff frequency of f_{co}=81.0 MHz.

All DUT RF emissions are collected at the sampling frequency rate, f_s=250 MSps using a near field probe with baseband bandwidth W_{bb}=500 MHz. The existing AFIT data set was collected using the LeCroy collection platform, and the ORNL dataset was collected using the National Instruments collection platform as shown in Table 3.1. The collected emissions are stored sequentially as 8 bit integer values representing the measured voltage level of the collected signal at evenly spaced time intervals.

For the National Instruments collection platform using Matlab®, two triggers are supplied to the collection platform to instantiate a signal collection. Both triggers must be present for a signal collection to occur. The first trigger is sent from a Laptop being used as an instrument controller, (controlling the RF-probe placement and the collection platform) to the collection platform indicating that the RF-probe is in position for a collection. This is done so that the physical movement of the RF-probe is synchronized with the collection platform and collections are not taken while the probe is moving between one of the N_L=100 locations. The second trigger is a threshold value, based on the voltage across a Light Emitting Diode (LED) on the DUT, prior to the first *MOV* operation of an LLP. The LED voltage is toggled as a square wave with an approximate duty cycle of 50% equal in length to the LLP scan time. This trigger indicated the start of an LLP scan.

When both triggers are present, indicating the probe is in position (first trigger) and an LLP scan has instantiated (second trigger) a t_{SIG}=5 ms collection is taken. A t_{SIG}=5 ms collection is taken to ensure the entire URE produced from the execution of an LLP scan is collected as the LLP scan is approximately t_{LLP}=3 ms. This triggering process is used for each N_L =100 locations, as well as the subsequent *refined probe placement* position.

3.3 Post Collection Processing

After signal collection, post-collection processing is done using Matlab®. The collections are read into Matlab® from the binary file and converted to type double for use with Matlab® filtering functions. The signals are then processed according to the following steps: 1) digital bandpass filtering 2) down-conversion to an intermediate frequency 3) down-sampling 4) SNR scaling.

1. *Bandpass Filtering* – The signals are bandpass filtered using a digital 8th-order Butterworth bandpass filter with a center frequency of f_{BP}=55.5 MHz and -3.0 dB bandwidth of W_{BP}=1.0 MHz. This is done using the built in Matlab® function `butter[]` to generate filter coefficients and `filtfilt[]` to perform the actual filtering. The magnitude response of the filter is shown in Figure 3.2.

2. *Downconversion* - After bandpass filtering, the signals are downconverted from the range of $f \in [55.0, 56.0]$ MHz to the range of $f \in [1.0, 3.0]$ MHz. Once

21

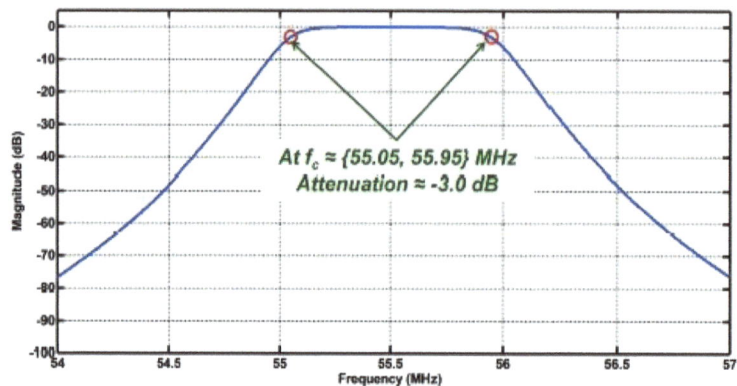

Figure 3.2: Magnitude Response of 8th order bandpass Butterworth Filter [43].

downconverted the signals are then digitally filtered with a LPF. The cutoff frequency for the LPF is $f_{co} = 3.5$ MHz.

3. *Downsampling* - After filtering and downconversion the signals are downsampled by a factor of $D_S 20$, reducing the number of samples to yield an effective sample rate of $f_s = 12.5$ MSps. Downsampling is accomplished by selecting the first element/sample of a signal, and henceforth every 20th element/sample, where unselected elements/samples are discarded.

4. *SNR Scaling* - independent Additive White Gaussian Noise (AWGN) realizations are added to the post-processed signals to simulate a range of channel Signal to Noise Ratio (SNR), . This is done to reduce the number of collections that would otherwise be needed to evaluate performance under degraded conditions. The range of SNR values presented in this research is, $SNR \in [-30 : 30]$ dB in $SNR_{step} = 5$ dB increments. For each signal collected, at each SNR considered, $N_{nz} = 10$ noise realizations are simulated. It is important to note that although *SNR scaling* is considered to be digital post-processing it occurs *after* ROI Extraction, discussed in the next section.

22

3.4 ROI Extraction

As mentioned previously an LLP execution takes approximately $t_{LLP}=3$ ms with the actual ROI spanning $t_{ROI}=1.5$ ms. The signal collection platform collects for $t_{sig}=5$ ms. This is done to ensure the entire ROI is captured in the collected waveform. ROI extraction process isolates the ROI from the unwanted part of the signal collection. ROI extraction occurs after digital signal processing, therefore all signal collections described henceforth are assumed to have been digitally post-processed according to Section 3.3 with the exception of added noise realizations.

Consider a given collection sequence $x_C[n] = x_C[1] + x_C[2]+\ldots+x_C[n]$, n=1,2...$N_C$, where N_C is the last collection sample. Also consider following sequences $x_{AS}[n]$ and $x_{ES}[n]$. $x_{AS}[n]$ represents the *alignment start*, i.e. discrete samples of the LLP operations {MOV,SQR} (those operations that begin every scan of the LLP that is being executed). $x_{ES}[n]$ represents the *alignment end*, the operations {SQR,MOV} (those operations that end each LLP).

The start of an ROI is determined by cross-correlating the collected signal sequence $x_C[n]$ with the *alignment start* sequence $x_{AS}[n]$. The end of an ROI is determined by cross-correlating $x_{ES}[n]$ with the signal collection $x_C[n]$. The ROI length is then estimated by finding the difference between lag values for the corresponding maximum cross-correlation values for the start time C_{MS} and end time C_{ME}.

The mean and standard deviation of the estimated ROI lengths are calculated, μ_{ROILen} and σ_{ROILen} respectively. A threshold value of $\mu_{ROILen} + 1.5\sigma_{ROILen}$ is established and any ROIs exceeding the threshold are discarded. As previously mentioned in Chapter 2 and shown in [43], unaccounted CPU operations can occur during the execution of the LLP. The extra CPU operations are unwanted and render an ROI unusable. ROIs that contain extra operations (those ROIs the exceed the length threshold) are therefore discarded.

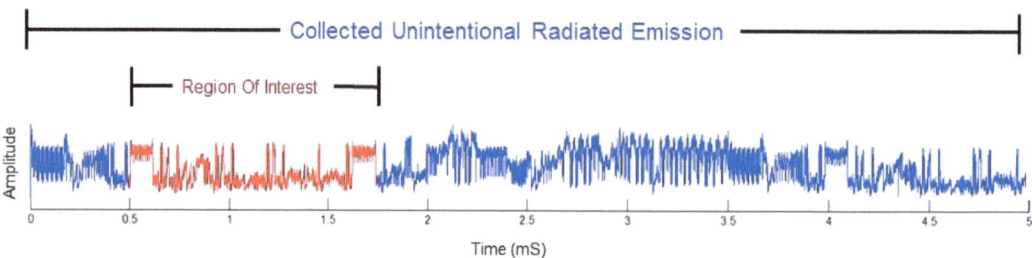

Figure 3.3: Region Of Interest Extraction

The remaining ROIs are sorted in descending order by the mean of their maximum cross-correlation values $\{C_{MS}, C_{ME}\}$. From this sorted set (first element in the set has the highest mean of $\{C_{MS}, C_{ME}\}$), the top 250 ROIs are chosen for Fingerprint generation. The remaining ROIs are not considered further. Signal collection at ORNL included multiple collections over the course of two consecutive days. Ultimately, 500 ROIs are used for fingerprint generation (250 ROIs from the two independent collections).

After selecting the best quality ROIs based on the correlation metrics described, the ROIs must be extracted from the collected signal. This is done by using the sample index of the maximum correlation start and end times.

3.5 Fingerprint Generation

RF-DNA fingerprint generation was implemented in accordance with previous AFIT RF-DNA fingerprinting research [4, 9, 36, 43]. The process has been applied to sequences representing Time-Domain (TD), Frequency Domain (FD) and Time-Frequency domain data sets [4, 9, 17, 36, 43, 49]. For the purpose of this research only TD features are considered.

For the complex signal $x[n] = x_{re}[n] + x_{im}[n]$, the instantaneous TD responses Amplitude $a[n]$, Phase $\phi[n]$ and Frequency $f[n]$ are given by,

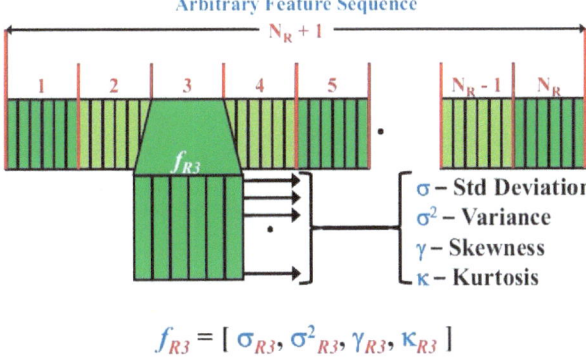

$$f_{R3} = [\; \sigma_{R3},\; \sigma^2{}_{R3},\; \gamma_{R3},\; \kappa_{R3}\;]$$

Figure 3.4: Abstract representation of RF-DNA fingerprint formation for an arbitrary sequence divided into N_R subregions [49]

.

$$a[n] = \sqrt{x_{re}[n]^2 + x_{im}[n]^2},\qquad(3.1)$$

$$\phi[n] = \tan^{-1}\left[\frac{x_{im}[n]}{x_{re}[n]}\right],\quad x_{re}[n] \neq 0,\qquad(3.2)$$

$$f[n] = \frac{1}{2\pi}\left[\frac{d\phi[n]}{dn}\right]\qquad(3.3)$$

The following steps give an overview of the RF-DNA fingerprinting process that is implemented for fingerprints generated from URE using the instantaneous TD features defined in Eqns. 3.1-3.3.

1. A selected ROI is divided into N_R equal contiguous time-domain sub-regions.

2. Within each subregion the mean μ is calculated and subtracted from all subregion samples to minimize the impact of collection bias.

3. The $N_{feat}=3$ instantaneous TD responses (Amp $a[n]$, Phz $\phi[n]$, Frq $f[n]$) are found for each subregion.

25

4. The $N_{stat}=4$ statistical attributes (standard deviation σ, variance σ^2, skewness γ, and kurtosis κ) are found for each subregion and each TD response.

5. The resulting statistical attributes are concatenated, and comprise an individual RF-DNA fingerprint that represents one ROI noise realization for a given SNR and given device.

6. The process is repeated for all ROIs across all noise realizations, all SNRs, and all devices.

In total there are $N_{\mathbf{F}} = 65000$ RF-DNA fingerprints per PLC device.

$$N_{\mathbf{F}} = N_{ROI} \times N_{nz} \times N_{SNR} \tag{3.4}$$

3.6 Feature Set Dimensional Reduction

The process described in Section 3.5 is implemented using $N_R=12$ subregions as well as calculating statistics over the entire ROI. The full dimensionality of a given fingerprint is therefore $N_D=156$.

$$N_D = (N_R + 1) \times N_{feat} \times N_{stat} \tag{3.5}$$

Reducing the fingerprint dimensionality is done by considering a subset of the full dimensional features. Qualitatively and Quantitatively selected subsets are considered. Table 3.3 details the dimensionality of the fingerprint sets used for classification and verification for the results presented in Chapter 4.

Of interest to this research is reducing the dimensionality of the feature set. Dimensional reduction is explored to enhance experimental-to-operational transition potential of RF-DNA fingerprinting [36].

Table 3.3: Feature sets used for Classification and Verification in GRLVQI & MDA/ML

Type	Feature Set	Number of Features	Number of Fingerprints
Full Dimensional	Full	156	500
Qualitative DRA	Amplitude	52	500
Qualitative DRA	Phase	52	500
Qualitative DRA	Frequency	52	500
Quantitative DRA	Top 33%	52	500
Quantitative DRA	Top 10%	16	500

3.6.1 Qualitative

Qualitative feature sets refer sets of fingerprints whose features are qualitatively selected and are solely composed of either Amplitude, Phase, or Frequency statistics. The same N_R subregions are used to calculate statistics, however the fingerprints in each qualitative set are of only one time domain response. The fingerprints are one third of the size of the full dimensional feature set fingerprints. The number of fingerprints in all feature sets is constant.

3.6.2 Quantitative

Features are ranked in descending order according to their relevance ranking determined by the GRLVQI process. Quantitative feature sets refer to fingerprints whose features are a subset of the full dimensional feature set that have been selected based on a relevance ranking. The Top 33% feature set is composed of the top ranked 52 of 156 features. Those features can be of any time domain response. Likewise the Top 10% feature set is composed of the top ranked 16 features. The Top 10% feature set is contained in the Top 33% feature set.

3.7 Device Discrimination

3.7.1 Classification

Following the formation of RF-DNA fingerprints, PLC hardware discrimination is performed. Two methods for PLC device classification are considered in this research, Generalized Relevance Learning from Vectors Quantized Improved (GRLVQI) and Multiple Discriminant Analysis Maximum Likelihood (MDA/ML). Although their respective internal mechanisms for hardware classification are different, the two independent processes make use of the same approach for both classification and verification. Both classification processes use RF-DNA fingerprints to identify a given PLC hardware device. RF-DNA fingerprints (generated as described in Section 3.5) are divided into two equal sized sets; Training fingerprints ($x_{TNG}[n]$) and Testing fingerprints ($x_{TST}[n]$). The fingerprints are divided based on an interleaved pattern (odd and even number indices). For the results shown in Chapter 4, the total number of fingerprints used is N_B=500 (N_{TNG}=250 and N_{TST}=250).

Training/Validation - The $x_{TNG}[n]$ set of fingerprints are used by the GRLVQI and MDA/ML processes to develop a device classification model. The set of $x_{TNG}[n]$ fingerprints are divided into k segments following a $k$fold partitioning process [36]. The GRLVQI and MDA/ML processes use k-1 segments to develop a classification model, where the kth segment is held out and is used after the model is developed to perform model-*validation*. The kth segment is introduced after model development to assess the performance of that model in correctly identifying a given devices fingerprints. All permutations of the k-fold model development and subsequent model-*validation* are carried out in turn. After all k-fold model development permutations, the model yielding the best validation results, (percent correct classification) is chosen. *Training/Validation* is repeated independently for each SNR considered.

Testing - Testing is the phase of model development where previously unseen fingerprints are introduced to the model that is selected during the validation phase. Testing assesses the model's ability to correctly identify devices. The $x_{TST}[n]$ set of fingerprints are held out of model development; $x_{TST}[n]$ fingerprints represent previously unseen data to assess the performance of the classification model.

3.7.2 Verification

Verification is a method of examination to determine how well a given device's $x_{TST}[n]$ fingerprints resemble what they are being classified as. Verification allows a one-to-one comparison based on a measure of similarity test statistic z_v. The test statistic used in both MDA/ML and GRLVQI is Euclidean Distance as derived in [9] [36] respectively.

A Probability Mass Function (PMF) of z_v is constructed for each device/class . A device's classified identity is then verified (correctly or incorrectly) by a binary decision against a threshold value t_v. If the test statistic is below the threshold the device is declared *Authorized* (correctly or incorrectly). Conversely a test statistic over the threshold is declared *Rogue*. The threshold value t_v as used in this research is implemented as described in [36].

$$x_{TST}[n] \rightarrow z_v[n] < t_v \quad : \quad Authorized \tag{3.6}$$

$$x_{TST}[n] \rightarrow z_v[n] > t_v \quad : \quad Rogue \tag{3.7}$$

3.8 Performance Evaluation

Classification performance as described in Section 2.3.1 is evaluated at an arbitrary baseline performance of 90% correct classification for a given SNR. The $C_{AVE}=90\%$ baseline performance metric has been used in previous AFIT research efforts [9, 36, 43].

The possible verification outcomes are shown in Table 3.4. Verification performance is assessed using Receiver Operating Characteristic (ROC) curve plots. There are two

Table 3.4: Authorized vs. Rogue Identification

Actual	Claimed	Declared	Outcome
Authorized	Authorized	Authorized	True Authorized Accept
Authorized	Authorized	Rogue	False Authorized Reject
Rogue	Authorized	Authorized	False Rogue Accept
Rogue	Authorized	Rogue	True Rogue Reject

types of ROC curves that are presented in Chapter 4, Authorized Device Identification and Rogue Device Rejection. Authorized Device Identification is a plot of False Verification Rate (FVR) vs. True Verification Rate (TVR). Rogue Device Rejection is a plot of Rogue Accept Rate (RAR) vs. TVR.

$$\text{FVR} = \frac{\sum FalseAuthorizedReject}{\sum TrueAuthorizedAccept + \sum FalseAuthorizedReject} \qquad (3.8)$$

$$\text{RAR} = \frac{\sum FalseRogueAccept}{\sum TrueRogueReject + \sum FalseRogueAccept} \qquad (3.9)$$

$$\text{TVR} = \frac{\sum TrueAuthorizedCount}{\sum TrueAuthorizedCount + \sum FalseAuthorizedCount} \qquad (3.10)$$

The Authorized ID plots show how much a given Authorized Device looks like itself when compared to the other Authorized Devices. The Rogue Device Identification plots shows how well a given classification model can correctly reject rogue devices and correctly accept authorized devices.

IV. Results

This chapter details the results of Programmable Logic Controller (PLC) device *classification* and *verification* processes using the Multiple Discriminant Analysis Maximum Likelihood (MDA/ML) and Generalized Relevance Learning from Vectors Quantized Improved (GRLVQI) classifiers as described in Section 2.3.1 using Radio Frequency Distinct Native Attribute (RF-DNA) fingerprints developed independently from the Lecroy and National Instruments (NI) collection platform data described in Table 3.1 are used as inputs for the *classification* and *verification* processes. Section 4.1 shows results for the Lecroy platform fingerprints, and Section 4.2 shows results for the NI platform fingerprints. For each collection platform there are 6 feature sets considered as listed in Table 3.3, where all Dimensional Reduction Analysis (DRA) Feature sets are subsets of the Full Dimensional feature set.

The RF-DNA fingerprints are developed from Unintentional Radiated Emission (URE) signal collections taken from $N_{DEV}=10$ PLC hardware devices where each device has $N_F=500$ fingerprints. As mentioned in Chapter 3 device ZC is not considered due to unpredictable device operation. The fingerprints are generated from Time-Domain (TD) signal responses as described in Section 3.5. Fingerprints in the range of Signal to Noise Ratio (SNR)$\in$[-30:5:30] dB are used for comparison of MDA/ML and GRLVQI classifiers as well as to compare the Lecroy and NI collection platforms. $N_{nz}=10$ independent Additive White Gaussian Noise (AWGN) realizations for each fingerprint at each SNR are used to simulate channel effects over the SNR range. For the purpose of comparing classifiers, feature sets, and collection platforms, *gain* is used to specify the difference of SNR at which respective performances are equivalent.

31

4.1 Expansion of Lecroy Platform RF-DNA Fingerprinting Results

The Lecroy collection platform data set was used in prior Air Force Institute of Technology (AFIT) research [7, 9, 43]. The existing data is used in this research to expand upon the previous device discrimination results i.e. results from the Lecroy collection platform shown, are from previously existing signal collections. The only new signal collections presented are signal collections taken using the NI collection platform. The expansion of Lecroy platform collection results includes analysis of feature dimensional reduction as well as the use of the MDA/ML classifier.

4.1.1 Full Dimensional

Figure 4.1(a)(b) shows *classification* results using the full dimensional feature set N_F=156, over the SNR range of [-30:30] dB in 5 dB increments. Figure 4.1(a) shows the MDA/ML classifier achieves a cross-device average C_{AVE}=90.0% correct classification for SNR>6.5 dB. Figure 4.1(b) shows the GRLVQI classifier achieves the C_{AVE}= 90.0% correct benchmark for SNR>11 dB. The MDA/ML classifier is outperforms the GRLVQI classifier with a G_{SNR}≈4.5 dB gain relative to the GRLVQI classifier.

4.1.2 Dimensional Reduction

Qualitative feature sets are dimensional reduced by using features generated from one of the N_{feat}=3 TD signal responses described in Section 3.5. The selected features are a subset of the full dimensional feature set where two TD signal response features types have been removed.

Consistent across the Lecroy platform Qualitative DRA feature sets, the MDA/ML classifier outperforms the GRLVQI classifier. The MDA/ML classifier has a G_{SNR}≈5 dB gain in the Amplitude set, a G_{SNR}≈3 dB gain in the Phase set, and G_{SNR}≈2.5 dB gain for the Frequency set against the respective GRLVQI Qualitative DRA feature set when comparing benchmark performance (90.0% correct classification).

32

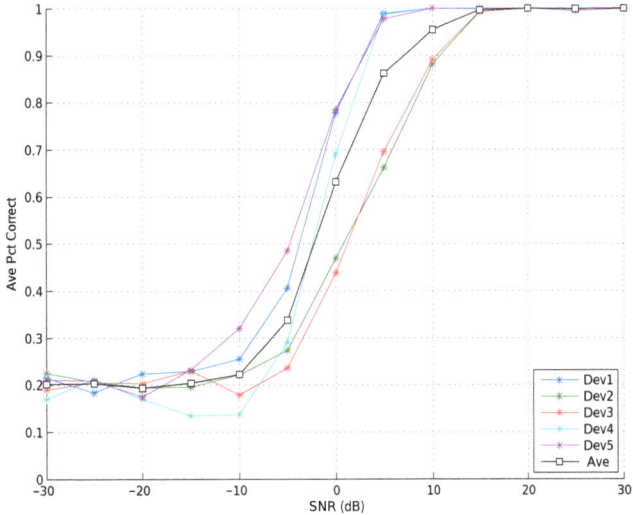

(a) MDA/ML Full feature Testing Results

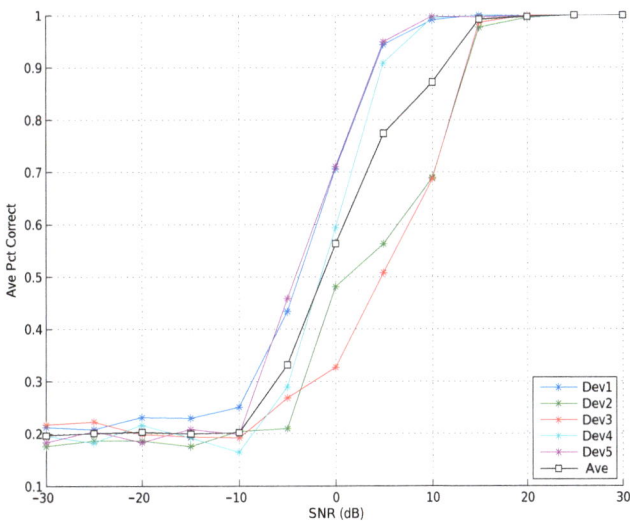

(b) GRLVQI Full Feature Testing Results

Figure 4.1: Full Dimensional Testing Results for MDA/ML and GRLVQI Lecroy Collection Platform

Of the three Qualitative DRA feature sets the Amplitude feature set yields best classification performance for both the MDA/ML and GRLVQI classifiers. Using the

33

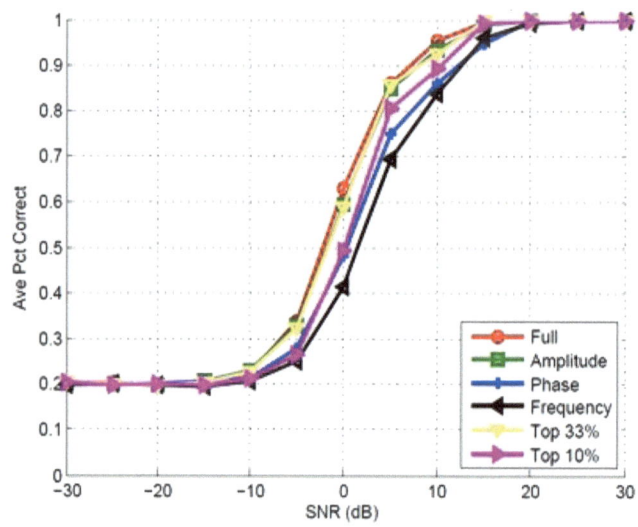

(a) MDA/ML Testing Results

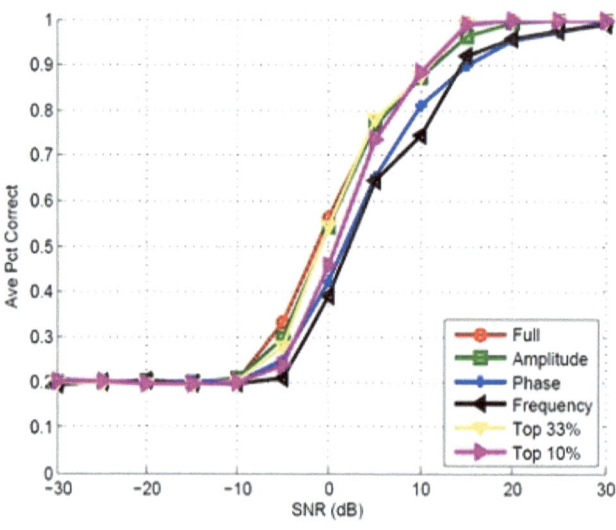

(b) GRLVQI Testing Results

Figure 4.2: Lecroy Platform DRA Testing Results

Amplitude Feature results in a benchmark performance gain of $G_{SNR} \approx 3$ dB when compared to both the Phase and Frequency DRA feature sets in both receiver platforms.

Quantitative feature sets, like the Qualitative feature sets, are subsets of the Full dimensional feature set. The Quantitative feature sets are constructed by examining the relevance rankings generated by the GRLVQI classifier. Features are assigned a weighted value (relevance rank) according to how well they impact *classification* performance [36]. There are two Quantitative feature sets, Top 33% and Top 10%. As previously mentioned both feature sets are contained in the Full dimensional feature set; the Top 10% feature set is a subset of the Top 33% feature set. The Quantitative feature sets are the respective percentage of the highest ranked features. Although the MDA/ML classifier does not have the inherent ability to produce relevance rankings, Quantitative feature sets constructed from the GRLVQI classifier are used by the MDA/ML. Fig. 4.3 shows an overlay of the relevance rankings for the Full dimensional feature set for both the Lecroy and NI collection platforms.

The Top 33% feature set outperformed the Top 10% feature set in MDA/ML *classification* by $G_{SNR} \approx 2$ dB. However for GRLVQI *classification* the Full, Top 10% and Top 33% feature sets are statistically equivalent using 95% confidence intervals.

4.2 National Instruments Platform RF-DNA Fingerprinting Results

4.2.1 Full Dimensional

Fig. 4.4 shows the *classification* results using the Full dimensional feature set $N_F = 156$ for collections taken with the NI platform over the SNR range of [-30:30] dB in $SNR_{Step} = 5$ dB increments. Fig. 4.4(a) shows the MDA/ML classifier achieves a cross-device average $C_{AVE} = 90.0\%$ correct classification for SNR>16.5 dB. Fig. 4.4(b) shows the GRLVQI classifier achieves the $C_{AVE} = 90.0\%$ correct benchmark for SNR>17 dB. The classifiers here achieve nearly the same performance.

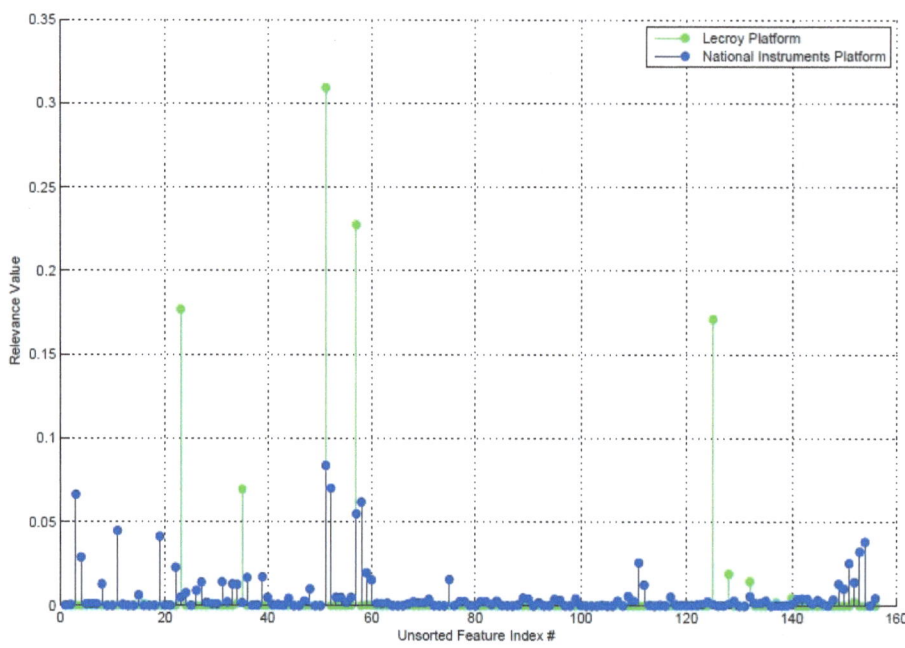

Figure 4.3: Relevance Rankings for Lecroy and NI Collection Platforms

4.2.2 Dimensional Reduction

Qualitative results using dimensionally reduced feature sets of N_F=52 are shown in Fig. 4.5(a)(b). Consistent across the NI Qualitative DRA feature sets, the GRLVQI classifier outperforms the MDA/ML classifier. This is opposite of classifier performance in Lecroy platform collection. The GRLVQI classifier has a G_{SNR}≈4 dB gain in the Amplitude set. The MDA/ML classifier does not meet the benchmark in the Phase feature set while the GRLVQI classifier reaches the benchmark for SNR>20 dB. The GRLVQI classifier sees a G_{SNR}≈2.5 dB gain in the Frequency set against the MDA/ML classifier.

The Amplitude Feature set yields best Qualitative DRA classification performance for each classifier. Using the Amplitude Feature set results in a benchmark performance gain of ≥7 dB when compared to both the Phase and Frequency DRA feature sets.

36

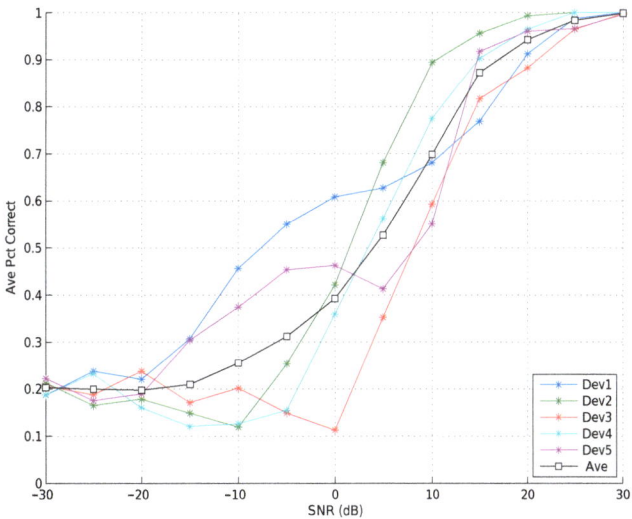

(a) MDA/ML Full Dimensional Testing Results

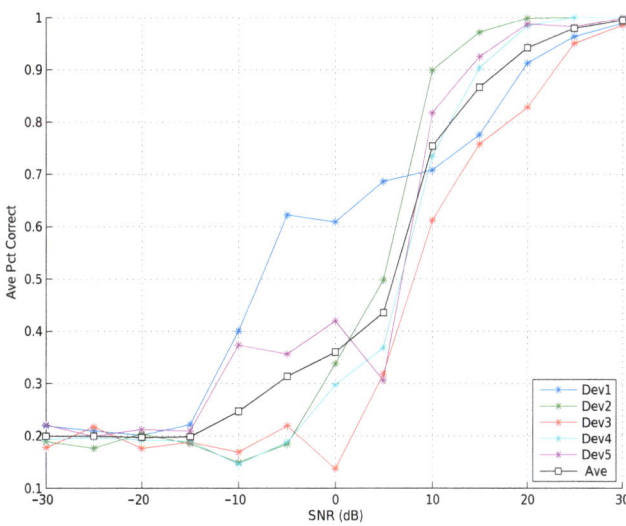

(b) GRLVQI Full Dimensional Testing Results

Figure 4.4: National Instruments Full Dimensional Testing Results

Quantitative The Top 33% feature set outperformed the Top 10% feature set in MDA/ML *classification* by $G_{SNR} \approx 7$ dB. The opposite is true for GRLVQI *classification* where the Top 10% feature set outperformed the Top 33% feature set by less than $G_{SNR} \approx 3$ dB. This is evident in both receiver platforms for the GRLVQI classifier. The

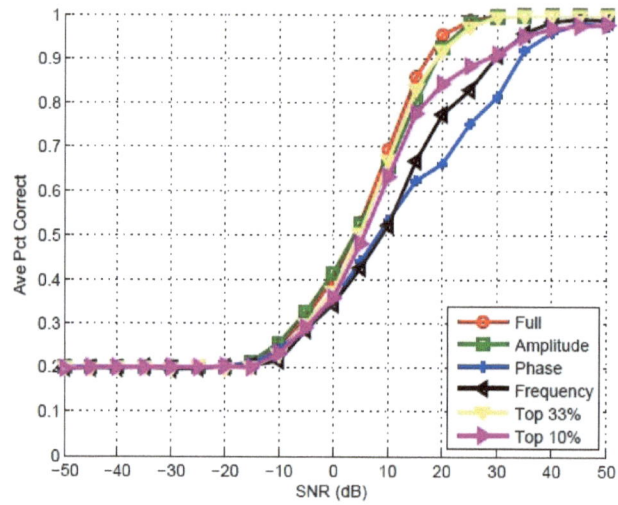

(a) MDA/ML Testing Results

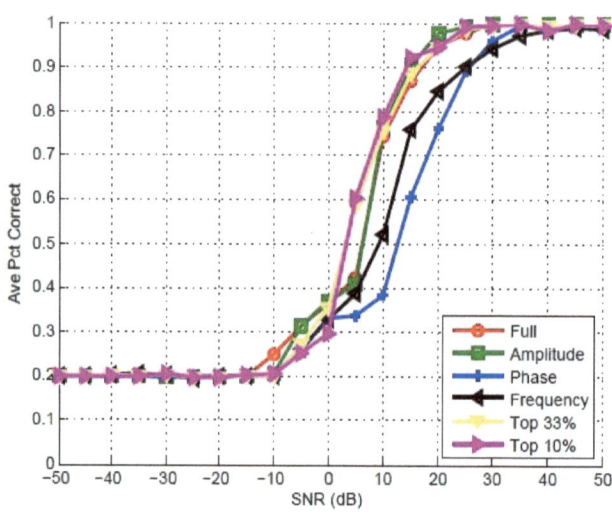

(b) GRLVQI Testing Results

Figure 4.5: National Instruments Platform DRA Testing Results

GRLVQI classifier is a Machine Learning Neural Network. Due to it's nature of model development, it is possible that the classifier suffers from *overlearning* [52] the characteristics for a given device's set of fingerprints. By using a model with less features, the model is more robust in characterizing device fingerprints in turn yielding better classification performance.

4.3 Device Verification

Device ID verification enables a one-to-one "how much like" device comparison. Results are shown for 5 Authorized devices $\{WQ, WV, KV, RG, OV\}$ and 4 Rogue devices $\{KG, QI, ZA, ZZ\}$. Devices were deemed Authorized or Rogue based upon their relative spectral emission intensity plots [43]. Receiver Operating Characteristic (ROC) curves are presented for the Full, and Top 10% feature sets for both MDA/ML and GRLVQI classifiers as well as both Lecroy and the NI collection platforms. The Full and Top 10% feature sets were chosen for presentation as they represent the extremes of feature dimensional reduction. In the interest of space, the other feature set *verification* plots are presented in the Appendix. The ROC curves are evaluated at the lowest value SNR where performance meets the arbitrary 90% average correct classification benchmark performance for the Full dimensional feature set. The corresponding SNR for the Lecroy collection platform is SNR=10 dB and the corresponding SNR for the NI platform is SNR=20 dB. *Verification* results are shown at these SNRs for the two collection platforms respectively. The test statistic used as a measure of similarity is Euclidean Distance for all ROC curves.

4.3.1 Authorized Device Identification

One aspect of *verification* is to verify the identity of a known authorized device. This is ability is assessed by comparing how similar the authorized devices resemble each other. The Equal Error Rate (EER) as described in Section 3.7.2 is used as the performance criteria. Results in Fig. 4.6(a)(b) shows results for both receiver platforms using the MDA/ML classifier. The MDA/ML classifier outperforms the GRLVQI classifier for both receiver platforms using the Full Dimensional set. All devices exceeded the EER of True Verification Rate (TVR)≥90% and False Verification Rate (FVR)≤10% for the Lecroy collection platform, 3 of 5 devices met the EER for the NI collection platform.

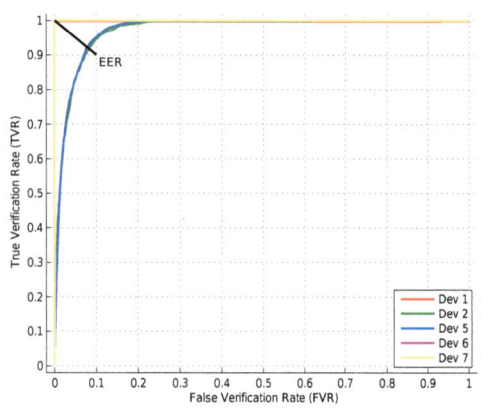

(a) Lecroy MDA/ML SNR=10 dB

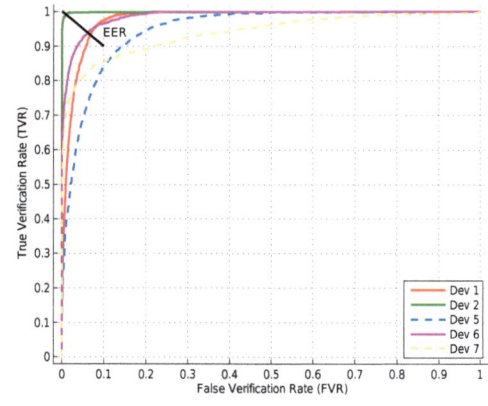

(b) National Instruments MDAML SNR=20 dB

Figure 4.6: Full Dimensional: Authorized ID Verification Results using MDA/ML

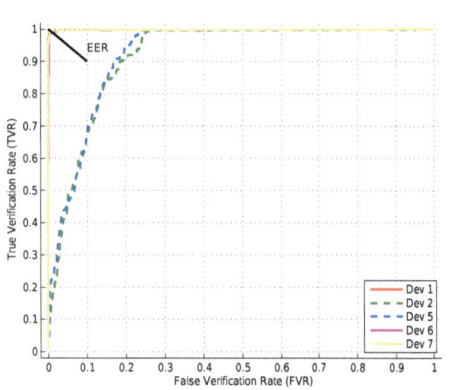

(a) Lecroy GRLVQI SNR=10 dB

(b) National Instruments GRLVQI SNR=20 dB

Figure 4.7: Full Dimensional: Authorized ID Verification Results using GRLVQI

Fig. 4.7(a)(b) shows results for both receiver platforms using the GRLVQI classifier, with 3 of 5 devices meeting the EER ≤10% benchmark for the Lecroy platform and 2 of 5 meeting the EER≤10% benchmark for the NI collection platform.

Fig. 4.8(a)(b) shows results for the Top 10% feature set using the MDA/ML classifier, and Fig. 4.9 shows results for the GRLVQI classifier. Device 5 does not meet the EER benchmark for any of the results presented using the Top 10% feature set for the verification of Authorized ID. Device 2 consistently fails meet the EER benchmark for both classifiers for the Lecroy collection platform.

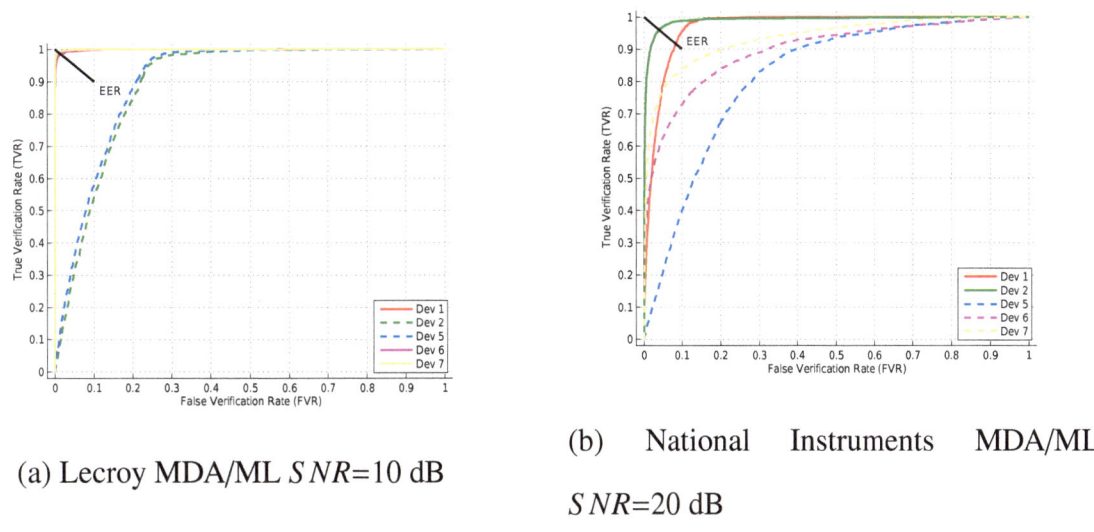

(a) Lecroy MDA/ML SNR=10 dB

(b) National Instruments MDA/ML SNR=20 dB

Figure 4.8: Top 10%: Authorized ID Verification Results using MDA/ML

4.3.2 Rogue Device Identification

For Rogue Device analysis the same thresholding procedure used to generate the ROC curves for *verification* of Authorized Device ID is used to generate ROC curves for Rogue Device Rejection as is implemented in Section 3.7.2. Correctly authorizing a known device is only one part of the device ID verification. Rogue devices, devices whose fingerprints

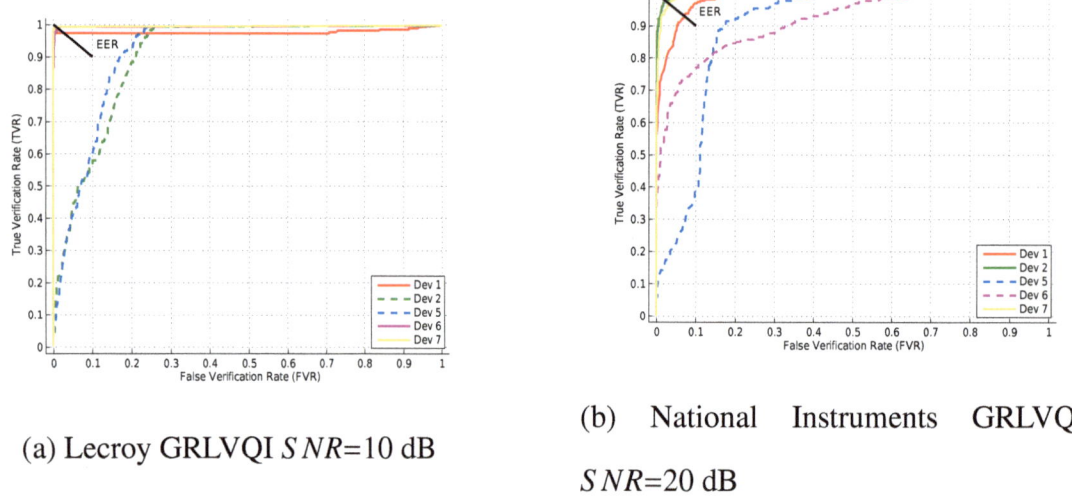

(a) Lecroy GRLVQI SNR=10 dB

(b) National Instruments GRLVQI SNR=20 dB

Figure 4.9: Top 10%: Authorized ID Verification Results using GRLVQI

have not yet been seen and are not represented in the classification models, must also be considered in verification solution. Rogue Device Identification measures "how much like" a given Rogue Device resembles each of the Authorized devices. This analysis simulates a Rogue device presenting bit-level credentials claiming to be a known authorized device and presenting itself for Device ID *verification*.

4.4 Cross Receiver Validation

By implementing the same collection process on two receiver collections platforms and allows direct comparison between the results. Although the Lecroy collection platform achieves the C_{AVE}=90% baseline performance with a gain of $G_{SNR} \approx 10$ dB over the NI platform, it should be noted that AWGN is added to the collected signals to degrade to meet the desired performance level. At the collected SNR (i.e. the absence of simulated AWGN) the NI collection platform is able to achieve results of 100% correct classification

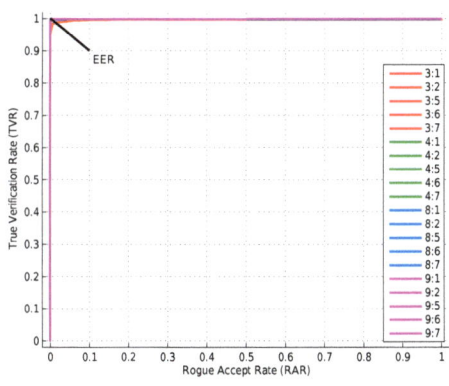

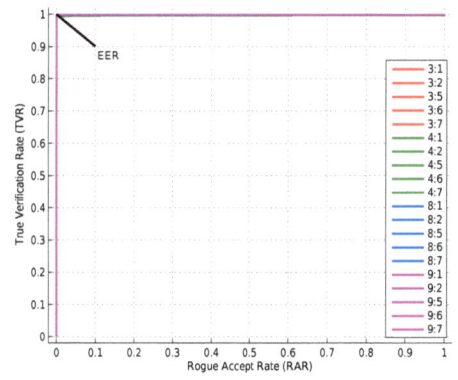

(a) Lecroy MDA/ML SNR=10 dB

(b) National Instruments MDA/ML SNR=20 dB

Figure 4.10: Full Dimensional: Rogue Device Rejection Verification Results using MDA/ML

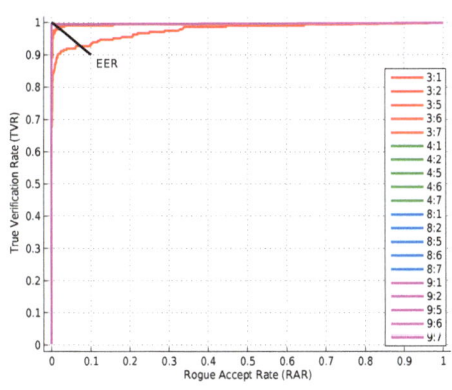

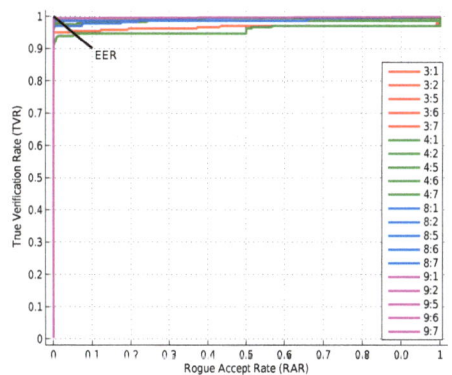

(a) Lecroy GRLVQI SNR=10 dB

(b) National Instruments GRLVQI SNR=20 dB

Figure 4.11: Full Dimensional: Rogue Device Rejection Verification Results using GRLVQI

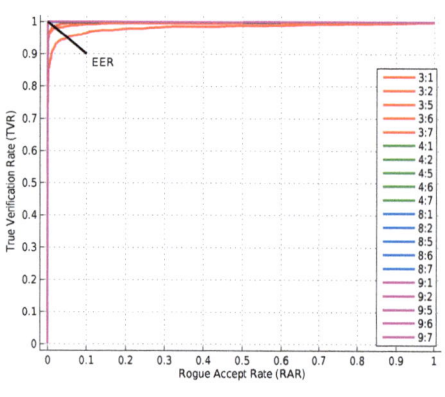
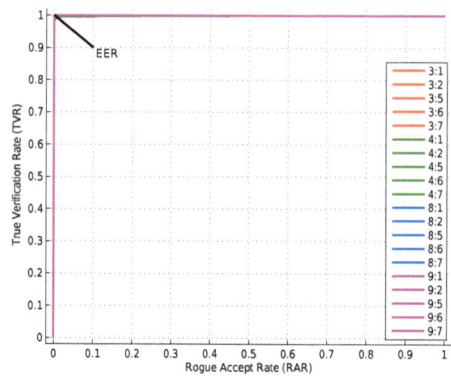

(a) Lecroy MDA/ML SNR=10 dB

(b) National Instruments MDA/ML SNR=20 dB

Figure 4.12: Top 10%: Rogue Device Rejection Verification Results using MDA/ML

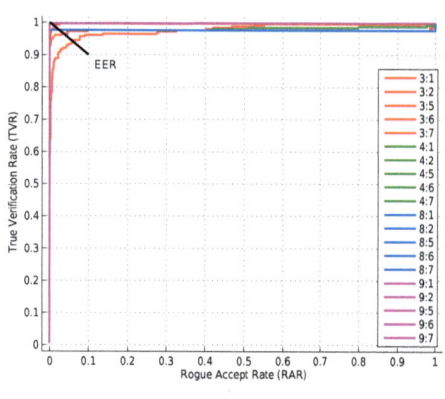
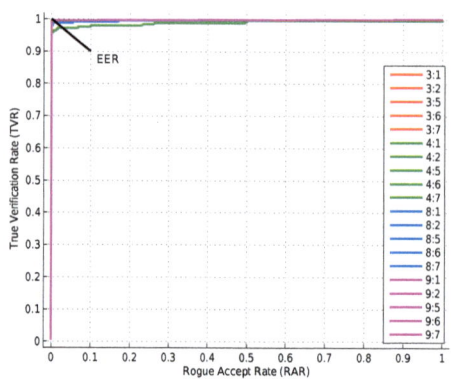

(a) Lecroy GRLVQI SNR=10 dB

(b) National Instruments GRLVQI SNR=20 dB

Figure 4.13: Top 10%: Rogue Device Rejection Verification Results using GRLVQI

for the Full dimensional feature set. Applying the verification process at the collected SNR, all devices meet the EER for both classifiers, for both Authorized Device ID Fig. 4.14 and Rogue Device Rejection Fig. 4.15.

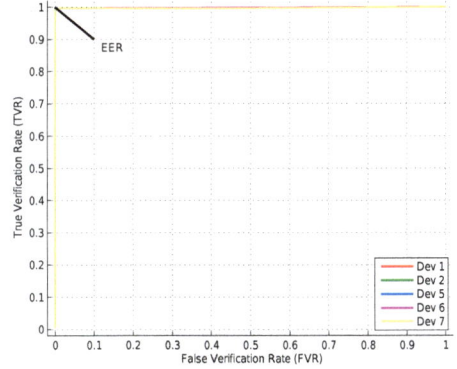

(a) National Instruments MDA/ML shown at the collected SNR

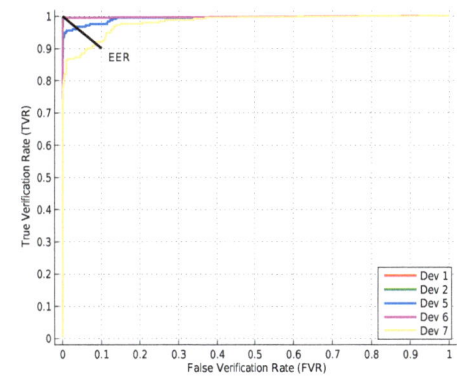

(b) National Instruments GRLVQI shown at the collected SNR

Figure 4.14: National Instruments Authorized ID results at the collected SNR

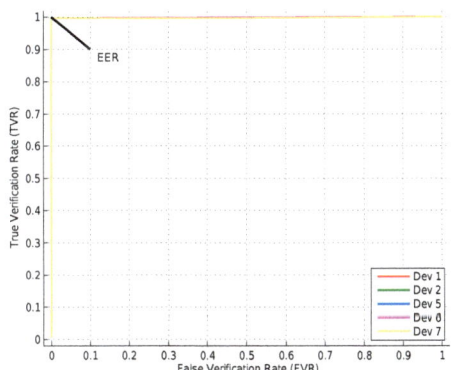

(a) National Instruments MDA/ML at the collected SNR

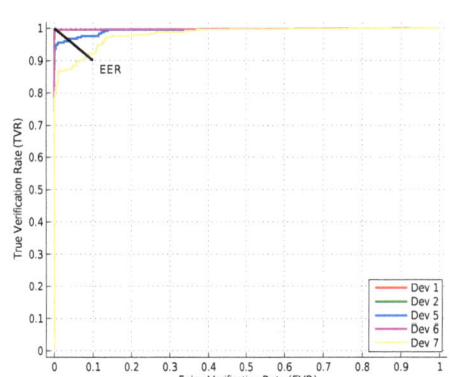

(b) National Instruments GRLVQI at the collected SNR

Figure 4.15: National Instruments Rogue Rejection results at the collected SNR

V. Conclusion

This chapter gives a summary of the results for Programmable Logic Controller (PLC) device discrimination using the Generalized Relevance Learning from Vectors Quantized Improved (GRLVQI) and Multiple Discriminant Analysis Maximum Likelihood (MDA/ML) classifiers considering dimensional reduction analysis of Radio Frequency Distinct Native Attribute (RF-DNA) time domain feature sets, using two different receiver platforms. Section 5.1 provides a summary of the key research activities. Sections 5.1.1-5.1.2 provide a summary of the research findings and results comparing classifier performance based on feature set dimensionality, as well as device discrimination performance based on receiver platform. Section 5.2 describes recommendations for future work of PLC device hardware discrimination and RF-DNA fingerprinting.

5.1 Research Summary

Improvement of cybersecurity in National Critical Infrastructure remains a government priority. Supervisory Control And Data Acquisition (SCADA) systems, which are used to control and monitor critical infrastructure such as waste water treatment centers, power generation plants, and traffic grids, are directly in line with this priority. PLCs are a basic unit of a SCADA system used to control low-level operations such as controlling the state of a valve, monitoring temperature or activating relays. As with almost all electronic devices PLCs make use of Integrated Circuits (IC)s which can be counterfeited or manufactured with hardware trojans [1, 10]. In critical SCADA applications potentially compromised hardware is a concern and could inflict grave damage. As such PLCs are chosen to demonstrate a proof of concept demonstration for a hardware device discrimination method.

Although much work has been done at securing PLCs at high layers of the Open Systems Interconnection (OSI) communication model, comparatively less research has been at the

46

lowest layer, the physical waveform layer. One method of augmenting higher layers of security by use of the physical layer is by exploiting characteristic differences in waveforms inherent to a particular device caused by component tolerances during manufacturing . This is one focus of Air Force Institute of Technology (AFIT)s Radio Frequency Intelligence (RFINT) program. The program has targeted many device and signal types with the goal of augmenting bit-level security, enabling human-like device discrimination and analyzing Side Channel Analysis vulnerabilities [4, 7, 9, 11–13, 15–20, 23, 25, 27, 38, 39, 43, 46, 49]

The goal of this research was to verify repeatability of existing AFIT signal collection methods for other receiver platforms, as well as to expand upon previous results by exploring the effects of fingerprint feature dimensional reduction. Verification of repeatability was accomplished by collecting Unintentional Radiated Emission (URE) from PLC devices in accordance with the collection procedure implemented in [43] using the National Instruments (NI) receiver platform at Oak Ridge National Laboratory (ORNL). Dimensional Reduction Analysis (DRA) was applied to the NI signal collection as well as previous data sets collected using the Lecroy collection platform used at AFIT. The results of these collections are shown in Chapter 4.

Additional research contributions were made by comparing the previously used MDA/ML and GRLVQI classifiers [4, 7, 9, 17, 36, 38, 49] using Time Domain RF-DNA fingerprints, to assess hardware component discrimination. The classifiers were used to perform classification of known authorized devices and verify their claimed identity, as well as detect and discriminant rogue devices.

Performance of *classification* was assessed using an arbitrary C_{AVE}=90% correct classification baseline performance as consistent with previous AFIT research [36]. *Verification* performance was assessed by 1) selecting the classification model with the lowest Signal to Noise ratio meeting the baseline performance 2) generating Receiver Operating Characteristic (ROC) curves at the associated Signal to Noise Ratio (SNR) and evaluating the Equal Error Rate (EER) point of True Verification Rate (TVR)$\geq$90% and False Verification Rate (FVR)$\leq$10%.

5.1.1 Cross-Platform Validation

To verify repeatability of signal collection methods, PLC device emissions were collected using an alternative receiver, the NI collection platform. The collection process was successfully repeated and comparable *classification* and *verification* results were obtained. All dimensional fingerprint sets considered in this research met the average C_{AVE}=90% correct classification baseline performance for both the MDA/ML and GRLVQI classifiers, albeit the sets achieved the baseline performance at varying SNRs. Although repeatable results were obtained the Lecroy platform outperformed the NI collection platform results by G_{SNR}=10dB in the Full Dimensional feature set for benchmark performance. The Full Dimensional baseline performance metric was used to select the SNR to evaluate ROC curves for the *verification* process.

The MDA/ML classifier performed best matched with the Lecroy receiver with a gain of $G_{SNR}\approx$3dB of performance averaged across feature sets versus the GRLVQI classifier. However the GRLVQI classifier when used with the NI collection platform saw a gain of $G_{SNR}\approx$4dB of baseline performance averaged across feature sets versus the MDA/ML classifier with a gain of G_{SNR}=12dB with the Top 10% feature set.

With the exception of one feature set for the Lecroy platform, both receivers failed to meet the EER for the *verification* of all authorized device IDs. Both receivers repeatedly failed to correctly verify device 5 regardless of feature set. However both receivers achieved 100% EER for Rogue Device Rejection.

5.1.2 Dimensional Reduction Analysis

Two types of feature dimensional reduction were considered in this research. Qualitative DRA feature sets were composed of statics generated from only one time-domain signal response {amplitude, phase, frequency}. Features in Quantitative DRA were selected based on GRLVQI relevance rankings. The GRLVQI has the inherent ability to rank features used in model development according to their influence in separating device/class vector representations. This provides an advantage over the MDA/ML classifier as the reduction of features improves memory storage, processing time, and classification processing complexity with acceptable loss in classification performance. Two Quantitative feature sets composed of a percentage of the top ranked features were considered: Top 33% and Top 10%. Both feature sets are subsets of the Full dimensional feature set, and the Top 10% feature set is contained in the Top 33% feature set.

MDA/ML performance is impaired by the reduction of feature dimensionality for all feature sets presented for *classification* and *verification* in both receivers. The opposite is true for the GRLVQI classifier in which the Top 10 % feature set matched or exceeded performance of the Full dimensional feature set based on 95% confidence intervals. Both receiver platforms ranked feature number 51 as the most influential feature for GRLVQI model development. Feature 51 is the Skewness of the Amplitude of Region 9.

5.2 Future Work Recommendations

The research results presented here show the effects of feature dimensional reduction in two different receiver platforms using two different classifiers. Both *classification* and *verificaiton* of PLC hardware device discrimination are shown to be succesful here and warrant continued investigation including,

1. Alternate RF-Probe: During the signal collection process outlined in 3.2.2 the placement of the RF probe requires precise alignment and any subsequent collections require the probe re-positioned. A less precise RF-probe, more akin to an antenna may not require such a rigorous placement routine and further mitigate challenges arising from repeatability.

49

2. Expansion of Feature Types: This research only considered Time-Domain signal response features. Previous AFIT research has shown other feature types such as Frequency Domain features and features derived from Gabor transforms to be successful for RF-DNA fingerprinting [4, 36].

3. Alternate IC Devices: Signal collections in this research were taken from the embedded microcontroller on the PLC mainboard. PLC device discrimination can be further expanded by using URE from other IC devices embedded on the mainboard to develop RF-DNA fingerprints.

4. Expansion of Software Anomaly Detection: Previous AFIT research assessed PLC ladder logic operation verification using Correlation Domain and Time Domain features [43]. PLC software anomaly detection can be further expanded by considering the feature dimensional reduction analysis demonstrated in this research.

VI. Appendix

This appendix presents are the remaining device *classification* results for both the Lecroy and National Instruments (NI) receiver platforms for both classifiers.

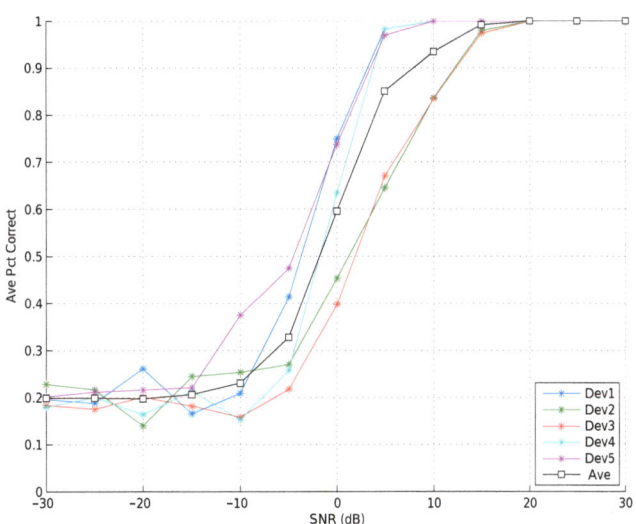

(a) Lecroy MDA/ML Amplitude Testing Results

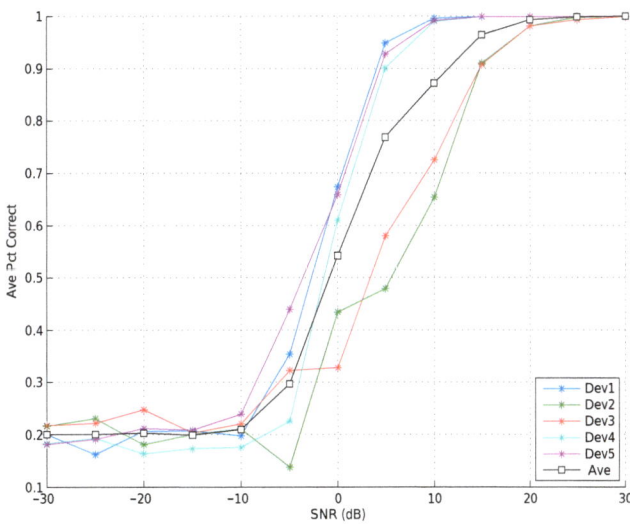

(b) Lecroy GRLVQI Amplitude Testing Results

Figure 6.1: Lecroy Qualitative Amplitude Classification Results

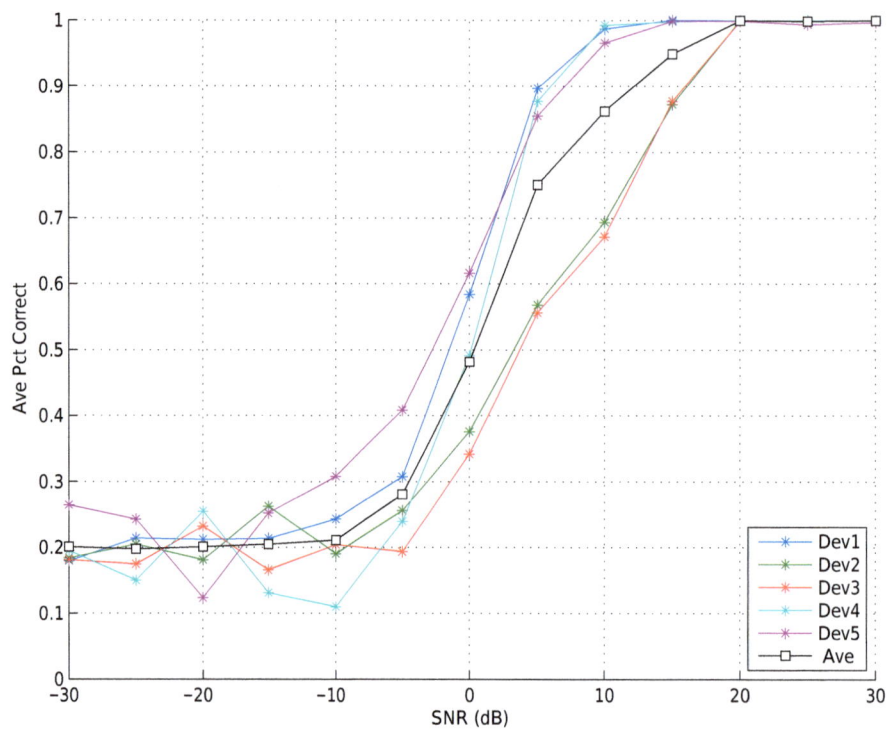

(a) Lecroy MDA/ML Phase Testing Results

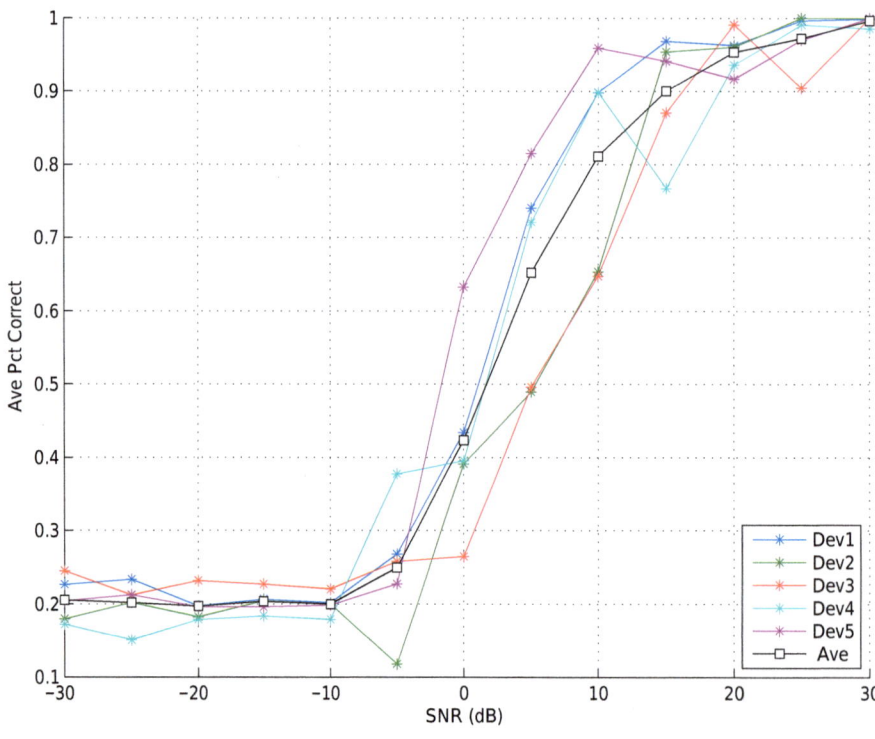

(b) Lecroy GRLVQI Phase Testing

Figure 6.2: Lecroy Qualitative Phase Classification Results

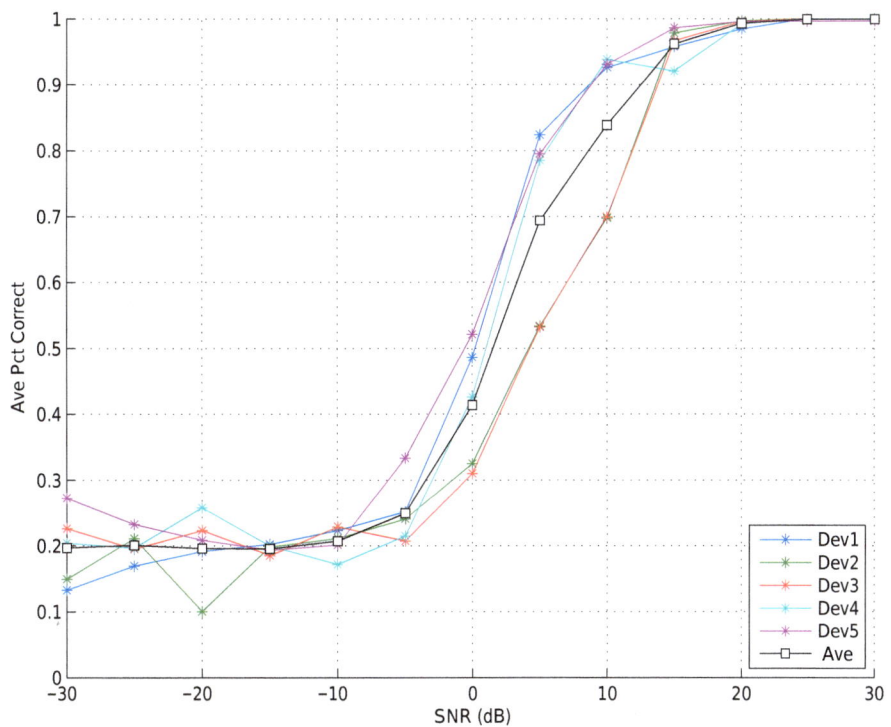

(a) Lecroy MDA/ML Frequency Testing Results

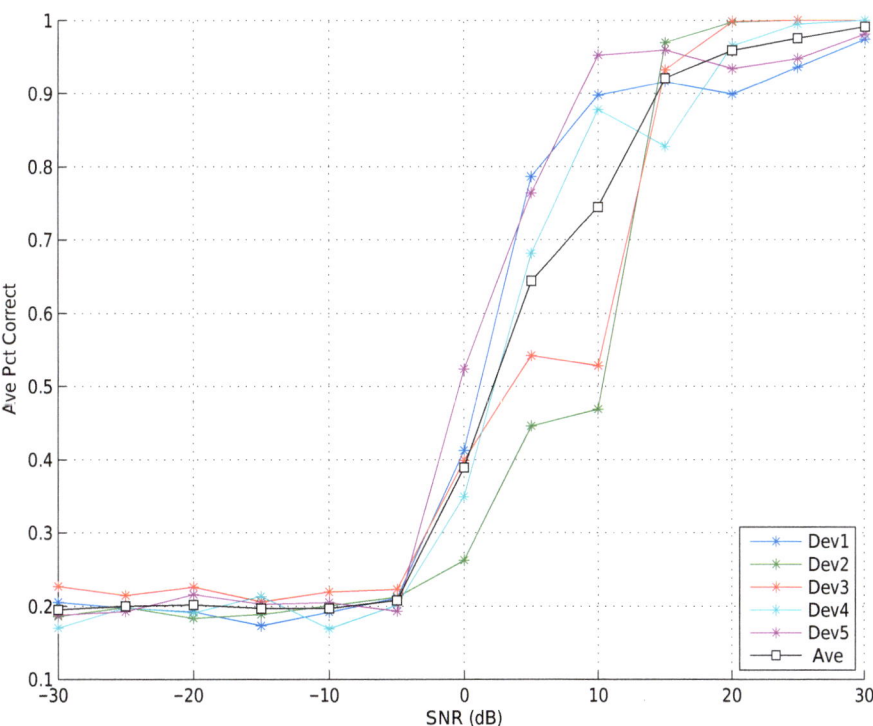

(b) Lecroy GRLVQI Frequency Testing Results

Figure 6.3: Lecroy Qualitative Frequency Classification Results

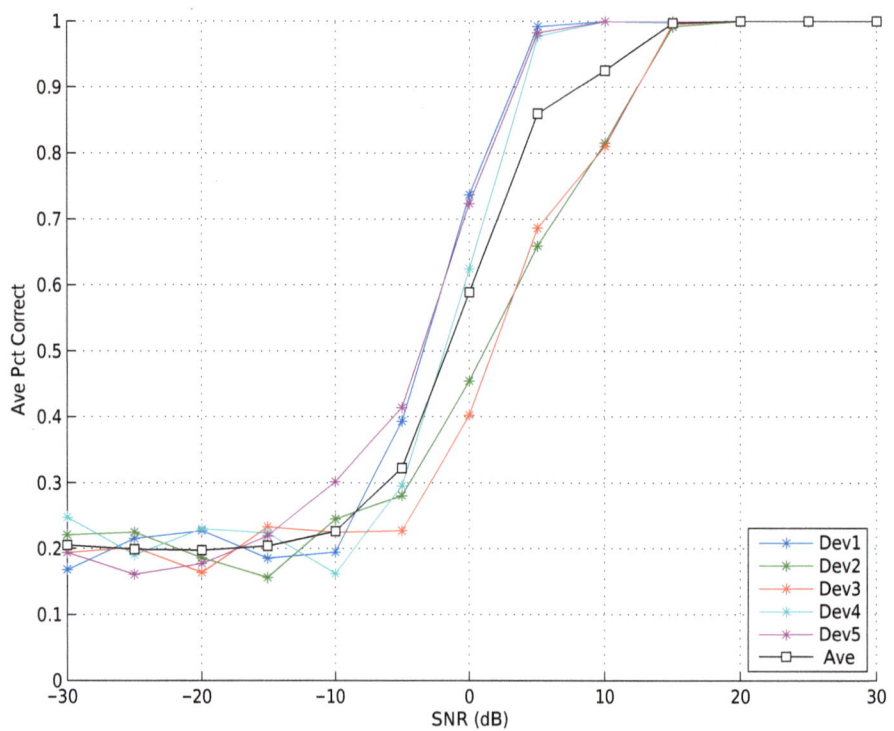

(a) Lecroy MDAML Top 33% Testing Results

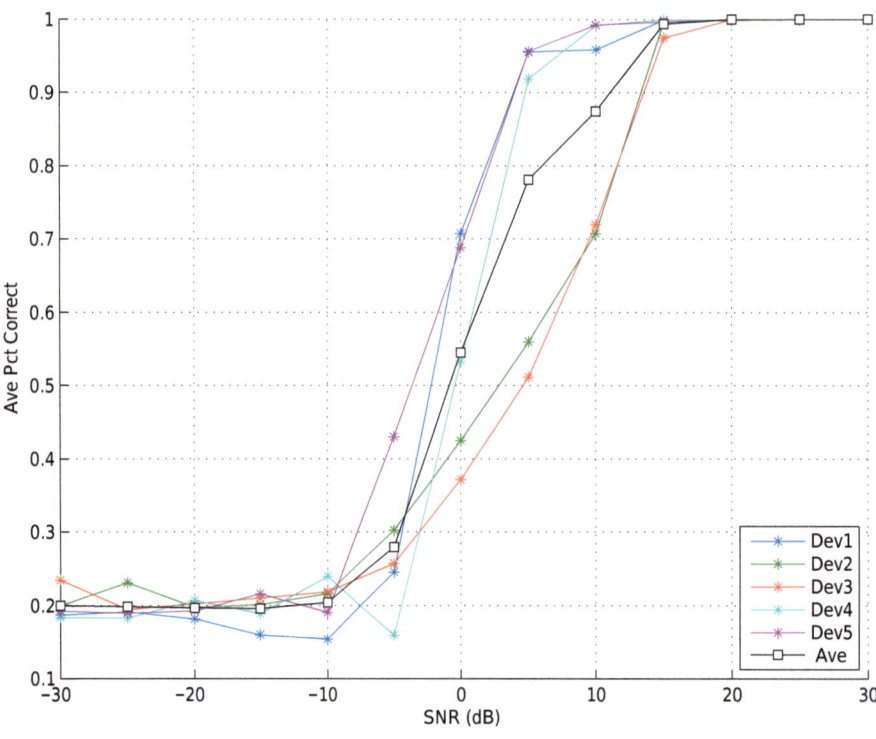

(b) Lecroy GRLVQI Top 33% Testing Results

Figure 6.4: Lecroy Quantitative Top33 Testing Averages

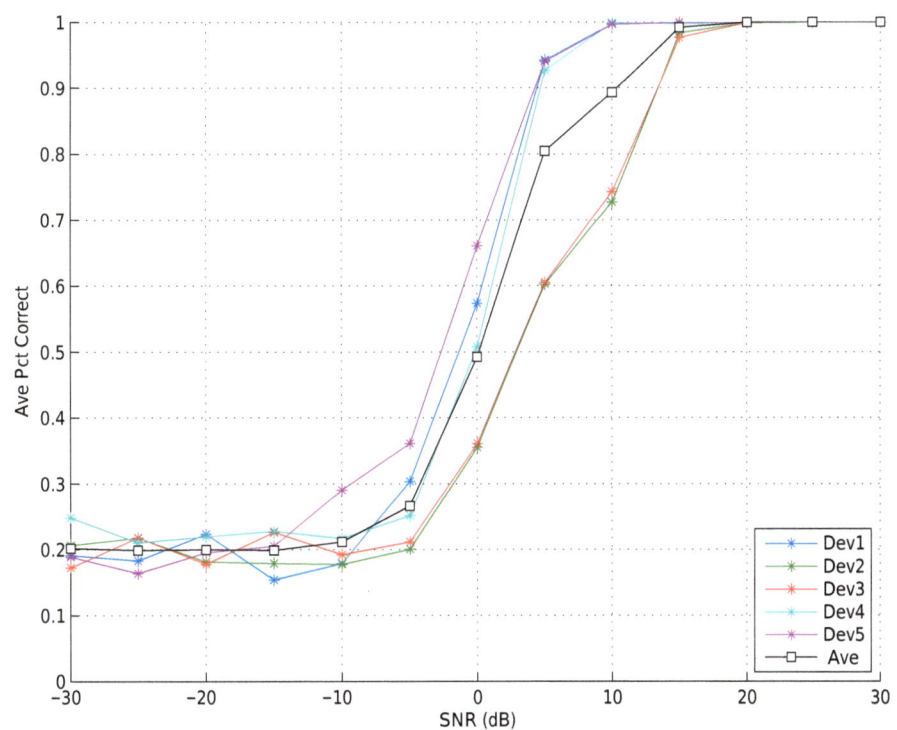

(a) Lecroy MDAML Top 10 % Testing Results

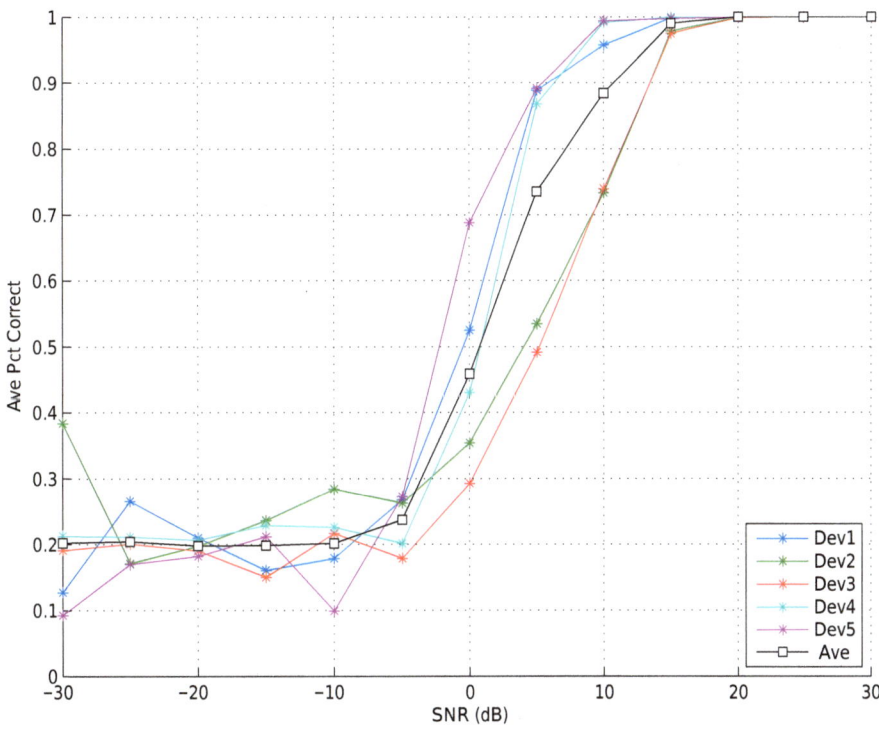

(b) Lecroy GRLVQI Top 10% Testing Results

Figure 6.5: Lecroy Quantitative Top 10% Testing Aves

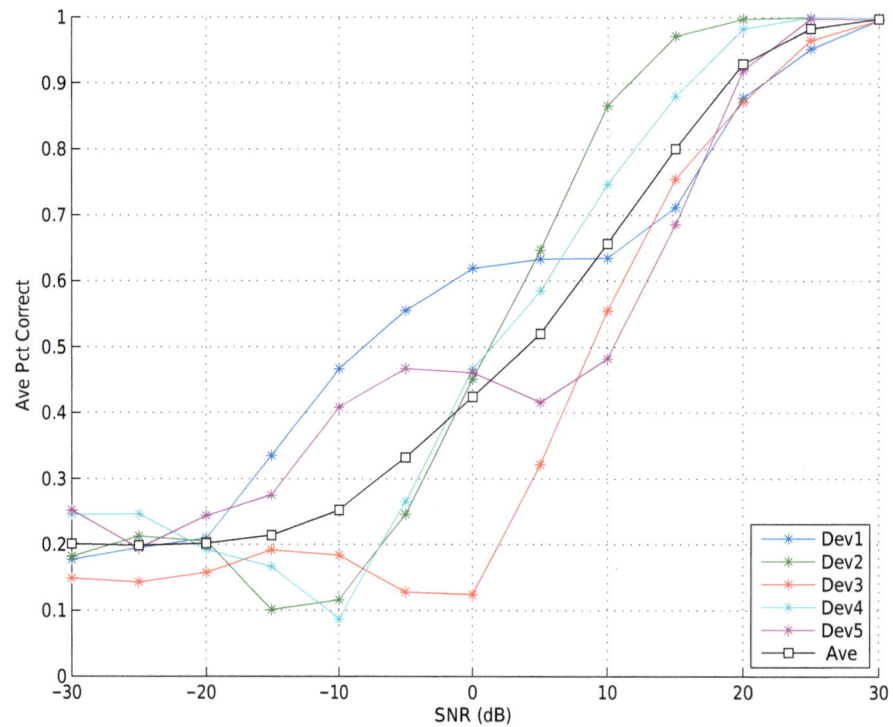

(a) NI MDAML Amplitude Testing

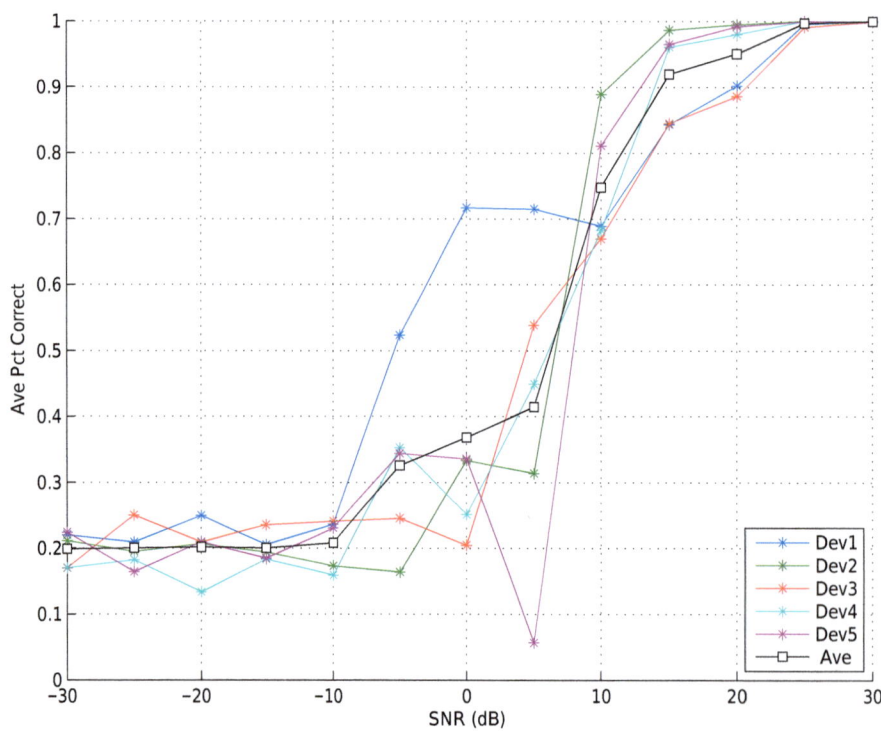

(b) NI GRLVQI Amplitude Testing

Figure 6.6: NI Amplitude Testing

56

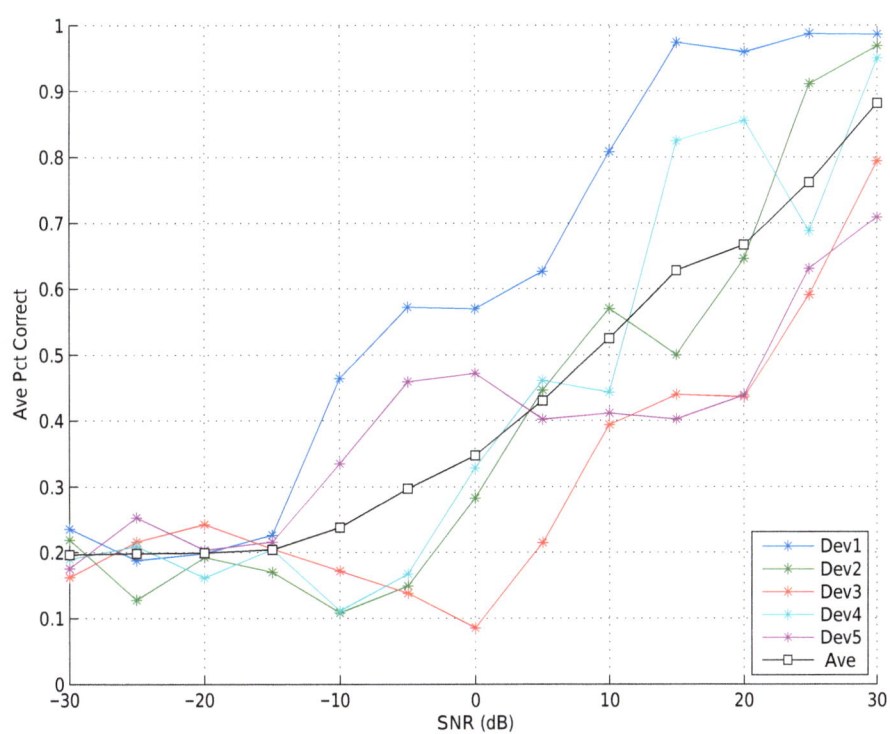

(a) NI MDAML Phase Testing

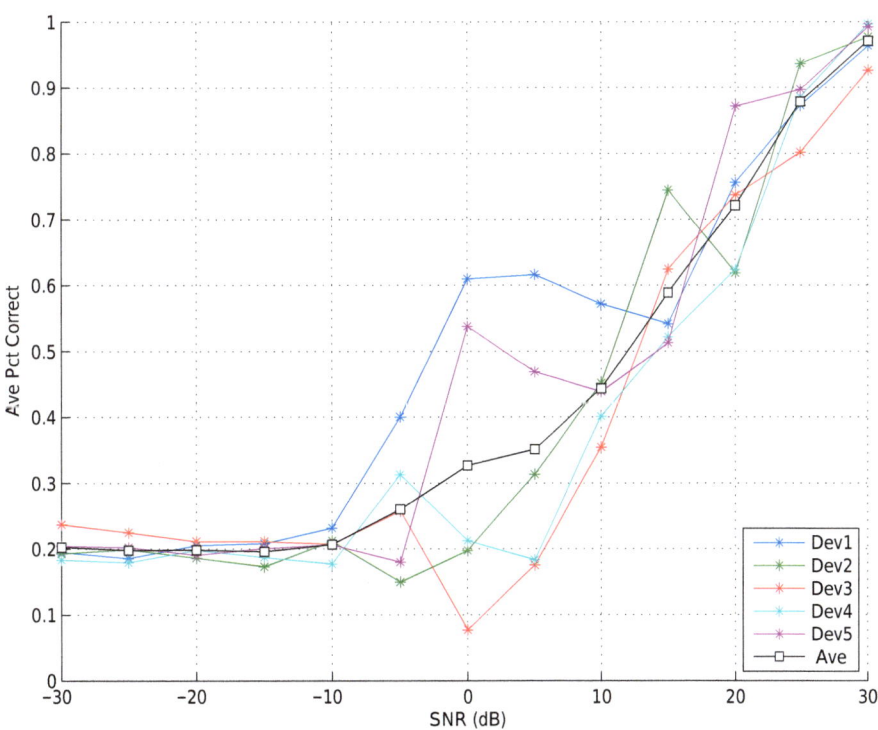

(b) NI GRLVQI Phase Testing

Figure 6.7: NI Phase Testing

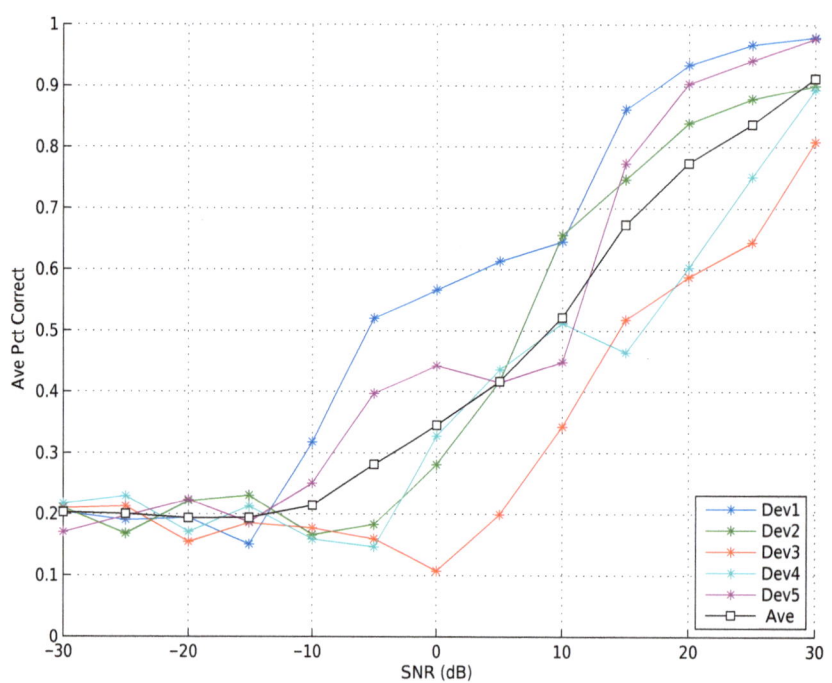

(a) NI MDAML Frequency Testing

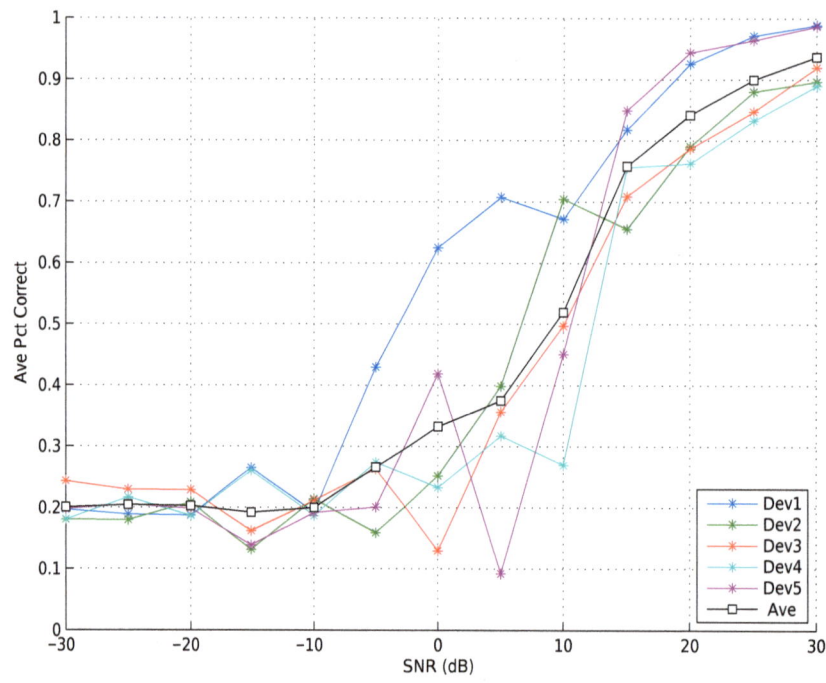

(b) NI GRLVQI Frequency Testing

Figure 6.8: NI Frequency Testing

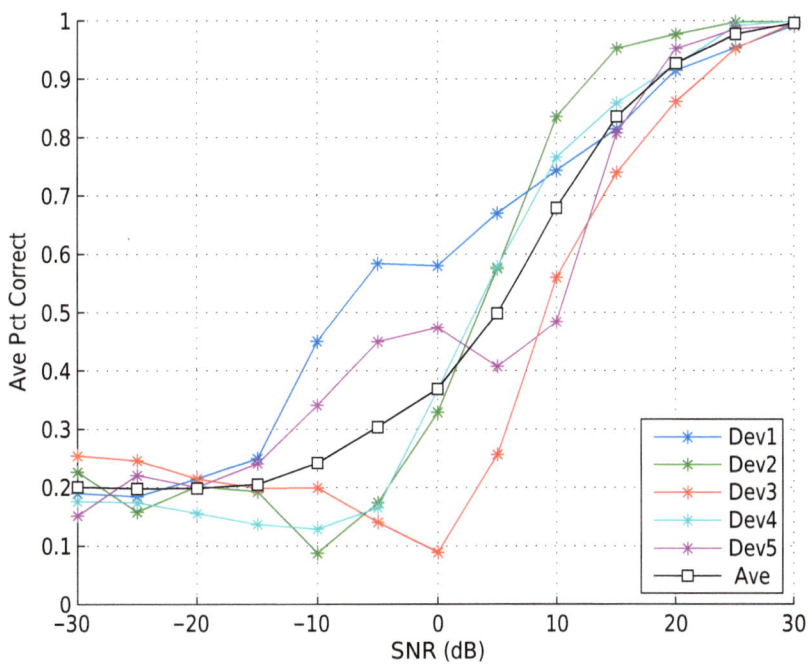

(a) NI MDAML Top 33% Testing

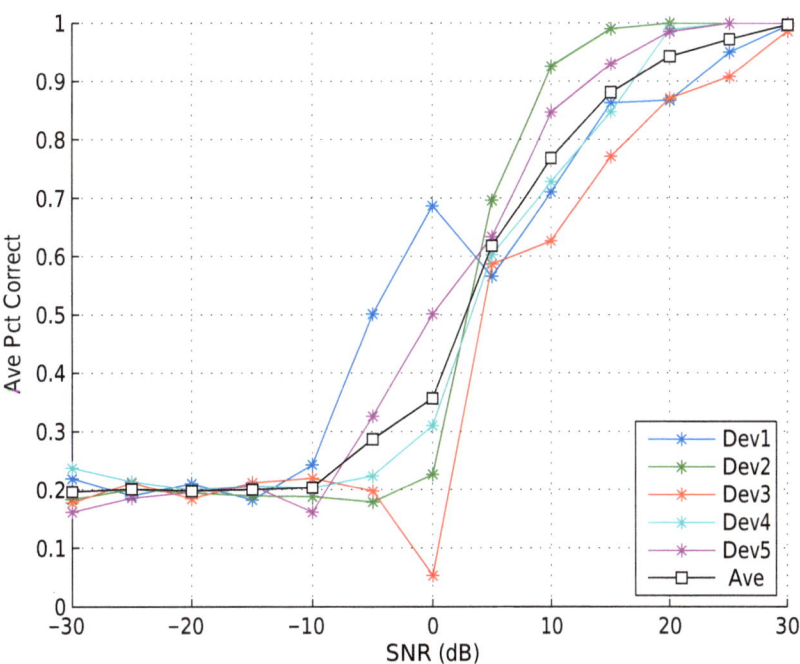

(b) NI GRLVQI Top 33% Testing

Figure 6.9: NI Top 33% Testing

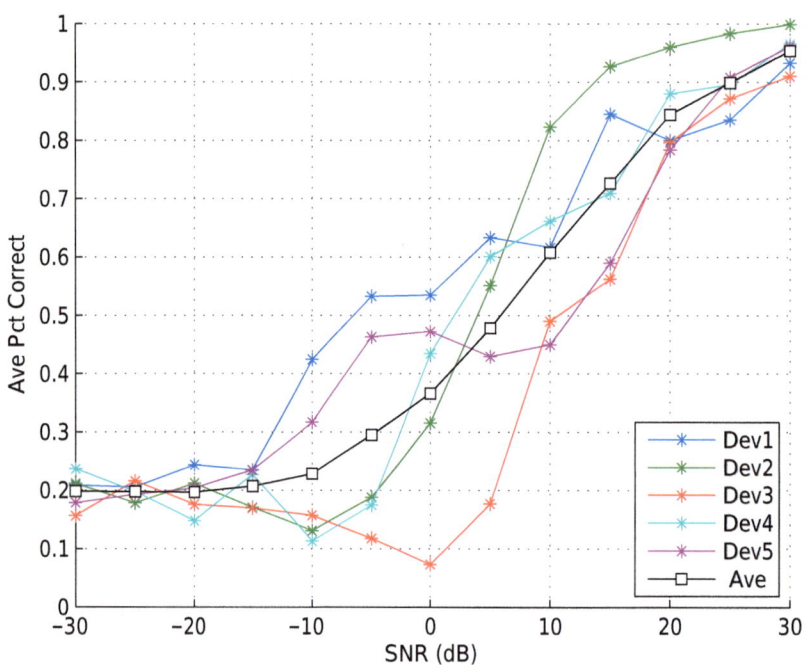

(a) NI MDAML Top 10% Testing

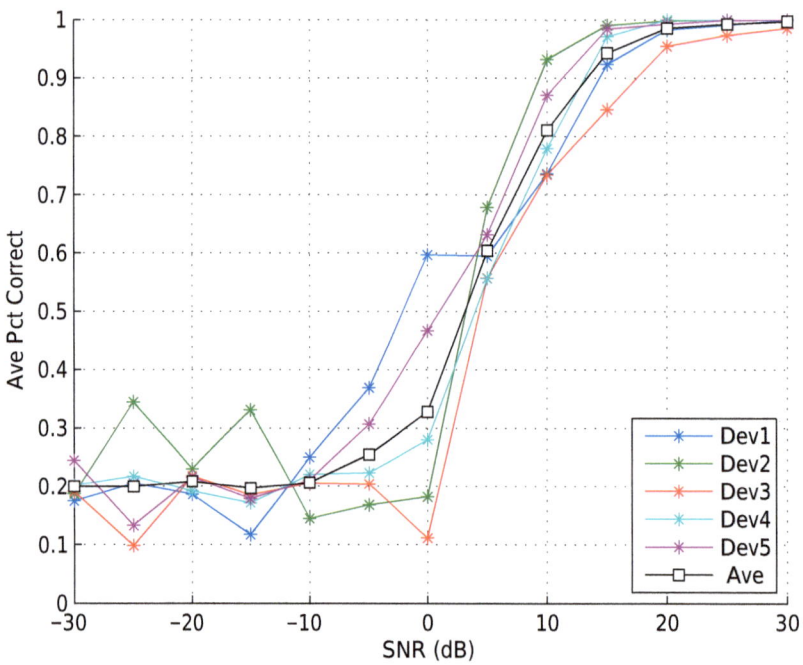

(b) NI GRLVQI Top 10% Testing

Figure 6.10: NI Top 10% Testing

60

Bibliography

[1] Adamov, A., A. Saprykin, D. Melnik, and O. Lukashenko. "The problem of Hardware Trojans detection in System-on-Chip". *CAD Systems in Microelectronics, 2009. CADSM 2009. 10th International Conference-The Experience of Designing and Application of,* 178–179. IEEE, 2009.

[2] Agrawal, D., S. Baktir, D. Karakoyunlu, P. Rohatgi, and B. Sunar. "Trojan detection using IC fingerprinting". *Security and Privacy, 2007. SP'07. IEEE Symposium on,* 296–310. IEEE, 2007.

[3] Alpaydin, E. *Introduction to machine learning.* MIT press, 2004.

[4] B. Hammer, B. and T. Villmann. "Generalized Relevance Learning Vector Quantization". *Neural Networks,,* 15:1059–1068. 2002.

[5] Ball, H. *USA Patriot Act of 2001.* ABC-CLIO, 2004.

[6] Cobb, W. *Exploitation of Unintentional Information Leakage from Integrated Circuits.* Technical report, DTIC Document, 2011.

[7] Cobb, W., E. Garcia, R. Baldwin, M. Temple, and Y. Kim. "Physical layer identification of embedded devices using RF-DNA fingerprinting". *Military Communications Conference,* 2168–2173. MILCOM 2010, 2010.

[8] Cobb, W., E. Laspe, R. Baldwin, M. Temple, and Y. Kim. "Intrinsic Physical Layer Authentication of ICs". *IEEE Trans on Information Forensics and Security,* 2(4):793–808, 2011.

[9] Cobb, W., E. Lsape, R. Baldwin, M. Temple, and Y. Kim. "Intrinsic Physical-Layer Authentication of Integrated circuits". *Information Forensics and Security,* volume vol. 7, 14–24. IEEE, February 2012.

[10] Collins, D. *Trust in Integrated Circuits.* Technical report, DTIC Document, 2008.

[11] Danev, B. and S. Capkun. "Transient-based identification of wireless sensor nodes". *Proceedings of the 2009 International Conference on Information Processing in Sensor Networks,* 25–36. IEEE Computer Society, 2009.

[12] Danev, B., T. Heydt-Benjamin, and S. Capkun. "Physical-layer Identification of RFID Devices." *Usenix Security Symposium,* 199–214. 2009.

[13] Danev, B., H. Luecken, S. Capkun, and K. El Defrawy. "Attacks on physical-layer identification". *Proceedings of the third ACM conference on Wireless network security,* 89–98. ACM, 2010.

61

[14] Das, S., K. Kant, and N. Zhang. *Handbook on Securing Cyber-Physical Critical Infrastructure*. Kaufmann, M, 2012.

[15] DeJean, G. and D. Kirovski. "RF-DNA: Radio-frequency certificates of authenticity". *Cryptographic Hardware and Embedded Systems-CHES 2007*, 346–363. Springer, 2007.

[16] Desmond, L., Tan Yuan, C., and R. Lee. "Identifying unique devices through wireless fingerprinting". *Proceedings of the first ACM conference on Wireless network security*, 46–55. ACM, 2008.

[17] Dubendorfer, C., B. Ramsey, and M. Temple. "An RF-DNA Verification Process for ZigBee Networks". *Military Communications Conference*, 1–6. MILCOM 2012, 2012.

[18] Dudczyk, J., J. Matuszewski, and M. Wnuk. "Applying the radiated emission to the specific emitter identification". *Microwaves, Radar and Wireless Communications, 2004. MIKON-2004. 15th International Conference on*, volume 2, 431–434. IEEE, 2004.

[19] Ellis, K. and N. Serinken. "Characteristics of radio transmitter fingerprints". *Radio Science*, 36(4):585–597, 2001.

[20] Gimelshteyn, M. *Classifying Commercial Receiver Emissions Using Fisher Discriminant Analysis*. Ph.D. thesis, Masters thesis, Air Force Institute of Technology, 2950 Hobson Way, WPAFB, OH, 2006.

[21] Gonzalez, C. and J. Reed. "Detecting unauthorized software execution in SDR using power fingerprinting". *MILITARY COMMUNICATIONS CONFERENCE, 2010-MILCOM 2010*, 2211–2216. IEEE, 2010.

[22] Hadley, M., m Lu, N, and A. Deborah. "Smart-grid security issues". *IEEE Security and Privacy*, 8(1):81–85, 2010.

[23] Hall, E., J. Budinger, R. Dimond, J. Wilson, and R. Apaza. "Aeronautical mobile airport communications system development status". *Integrated Communications Navigation and Surveillance Conference (ICNS), 2010*, A4–1. IEEE, 2010.

[24] Hall, M. *Correlation-based feature selection for machine learning*. Ph.D. thesis, The University of Waikato, 1999.

[25] Harmer, P. and M. Temple. "An improved LFS engine for physical layer security augmentation in cognitive networks". *Computing, Networking and Communications (ICNC), 2013 International Conference on*, 719–723. IEEE, 2013.

[26] Harmer, P., M. Temple, M. Buckner, and E. Farquhar. "4G Security Using Physical Layer RF-DNA with DE-Optimized LFS Classification." *Journal of Communications*, 6(9), 2011.

[27] Harmer, P., M. Williams, and M. Temple. "Using DE-Optimized LFS Processing to Enhance 4G Communication Security". *Computer Communications and Networks (ICCCN), 2011 Proceedings of 20th International Conference on*, 1–8. IEEE, 2011.

[28] Harmer, P., MD Williams, and M. Temple. "Using DE-Optimized LFS Processing to Enhance 4G Communication Security". *Computer Communications and Networks (ICCCN), 2011 Proceedings of 20th International Conference on*, 1–8. IEEE, 2011.

[29] Keller, W. and B. Pathak. "Integrated circuit with electromagnetic energy anomaly detection and processing", Mar, publisher=Google Patents, note=US Patent App. 13/410,909 2012.

[30] Klein, R. *Application of Dual-Tree Complex Wavelet Transforms to Burst Detection and RF Fingerprint Classification*. Dissertation, Air Force Institute of Technology, 2009.

[31] Klein, R., M. Temple, and and Reising D. Mendenhall, M. "Sensitivity analysis of burst detection and RF fingerprinting classification performance". *Communications, 2009. ICC'09. IEEE International Conference on*, 1–5. IEEE, 2009.

[32] Klein, R., M. Temple, and M. Mendenhall. "Application of wavelet-based RF fingerprinting to enhance wireless network security". *Communications and Networks, Journal of*, 11(6):544–555, 2009.

[33] Mateti, P. "Hacking Techniques in Wireless Networks". *Handbook of Information Security, Threats, Vulnerabilities, Prevention, Detection and Management*, 3:83, 2006.

[34] Mendenhall, M. and E. Merenyi. "Relevance-Based Feature Extraction for Hyperspectral Images". *Neural Networks, IEEE Trans on,*, 19(4):658–672. IEEE, April 2008.

[35] Petruzella, F. *Programmable Logic Controllers*. McGraw-Hill, fourth edition, 2005.

[36] Reising, D. *Exploitation of RF-DNA for Device Classification and Verification Using GRLVQI Processing*. Dissertation, Air Force Institute of Technology, 2012.

[37] Reising, D. and M. Temple. "WiMAX mobile subscriber verification using Gabor-based RF-DNA fingerprints". *Communications (ICC), 2012 IEEE International Conference on*, 1005–1010. IEEE, 2012.

[38] Reising, D., M. Temple, and M. Mendenhall. "Improving intra-cellular security using air monitoring with RF fingerprints". *Wireless Communications and Networking Conference (WCNC), 2010 IEEE*, 1–6. IEEE, 2010.

[39] Reising, D., M. Temple, and M. Oxley. "Gabor-based RF-DNA fingerprinting for classifying 802.16 e WiMAX mobile subscribers". *Computing, Networking and Communications (ICNC), 2012 International Conference on*, 7–13. IEEE, 2012.

63

[40] Shaw, W. *Cybersecurity for SCADA systems*. Pennwell books, 2006.

[41] Skanata, D and D. Byrd. "An overview of energy and telecommunications interdependencies modeling at NISAC". *Computational Models of Risks to Infrastructure*, 13:36, 2007.

[42] Sklar, B. *Digital communications*, volume 1099. Prentice Hall PTR New Jersey, 2001.

[43] Stone, S. *Radio Frequency Based Programmable Logic Controller Anomaly Detection*. Dissertation, Air Force Institute of Technology, 2013.

[44] Stone, S. and M. Temple. "Radio-frequency-based anomaly detection for programmable logic controllers in the critical infrastructure". *International Journal of Critical Infrastructure Protection*, 5(2):66–73, 2012.

[45] Stone, S., M. Temple, and R. Baldwin. "Detecting Anomalous SCADA Operation Using RF-Based Hilbert Transforms." *International Journal of Critical Infrastructure Protection*, 5(2), 2013.

[46] Suski, W., M. Temple, M. Mendenhall, and R. Mills. "Radio frequency fingerprinting commercial communication devices to enhance electronic security". *International Journal of Electronic Security and Digital Forensics*, 1(3):301–322, 2008.

[47] Suski, W., M. Temple, M. Mendenhall, and R. Mills. "Using spectral fingerprints to improve wireless network security". *Global Telecommunications Conference, 2008. IEEE GLOBECOM 2008. IEEE*, 1–5. IEEE, 2008.

[48] V., Vipendra. "Concept of Open Systems Interconnection (OSI) Model in Networking". March 2012.

[49] Williams, M., S. Munns, M. Temple., and M. Mendenhall. "RF-DNA Fingerprinting for Airport WiMax Communications Security". Number pp. 32-39 in 4th International Conference. Network and System Security (NSS), Sep. 2012.

[50] Williams, MD., M. Temple, and D. Reising. "Augmenting bit-level network security using physical layer RF-DNA fingerprinting". *Global Telecommunications Conference (GLOBECOM 2010), 2010 IEEE*, 1–6. IEEE, 2010.

[51] Wright, J. "Detecting wireless LAN MAC address spoofing". *White Paper, January*, 2003.

[52] Yegnanarayana, B. *Artificial neural networks*. PHI Learning Pvt. Ltd., 2009.

[53] Zetter, K. "How digital detectives deciphered Stuxnet, the most menacing malware in history". *Retrieved February*, 1:2012, 2011.